COHOMOLOGY OPERATIONS

LECTURES BY

N. E. Steenrod

WRITTEN AND REVISED BY

D. B. A. Epstein

ANNALS OF MATHEMATICS STUDIES

PRINCETON UNIVERSITY PRESS

A

Annals of Mathematics Studies

Number 50

COHOMOLOGY OPERATIONS

LECTURES BY

N. E. Steenrod

WRITTEN AND REVISED BY

D. B. A. Epstein

PRINCETON, NEW JERSEY

PRINCETON UNIVERSITY PRESS

1962

L.C. Card 62-19961
ISBN 0-691-07924-2

Fourth Printing, 1974

Printed in the United States of America
by Princeton University Press, Princeton, New Jersey

PREFACE

Speaking roughly, cohomology operations are algebraic operations on the cohomology groups of spaces which commute with the homomorphisms induced by continuous mappings. They are used to decide questions about the existence of continuous mappings which cannot be settled by examining cohomology groups alone.

For example, the extension problem is basic in topology. If spaces X, Y, a subspace $A \subset X$, and a mapping $h: A \longrightarrow Y$ are given, then the problem is to decide whether h is extendable to a mapping $f: X \longrightarrow Y$. The problem can be represented by the diagram

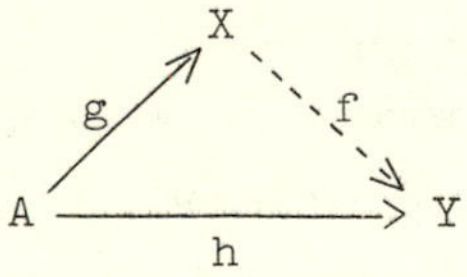

$$fg = h$$

where g is the inclusion mapping. Passing to cohomology yields an algebraic problem

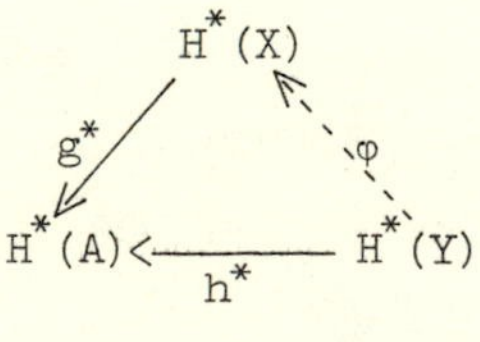

$$g^*\varphi = h^*$$

If f exists, then $\varphi = f^*$ solves the algebraic problem. In general the algebraic problem is weaker than the geometric problem. However the more algebraic structure which we can cram into the cohomology groups, and which φ must preserve, the more nearly will the algebraic problem approximate the geometric. For example, φ is not only an additive homomorphism of groups, but must be a homomorphism of the ring structures based on the cup product. Even more, φ must commute with all cohomology operations.

In these lectures, we present the reduced power operations (the squares Sq^i and p^{th} powers P^i where $i = 0,1,\ldots$, and p is a prime). These are constructed, and their main properties are derived in Chapters V, VII and VIII. These chapters are independent of the others and may be read first. Chapter I presents the squares axiomatically, all of their main properties are assumed. In Chapters II, III, and IV, further properties are developed, and the principal applications are made. Chapter VI contains axioms for the P^i $(p > 2)$, and applications of these. Chapter VIII contains a proof that the squares and p^{th} powers are characterized by some of the axioms assumed in I and VI.

The method of constructing the reduced powers, given in VII, is new and, we believe, more perspicuous. The derivation of Adem's relations in VIII is considerably simpler than the published version. The uniqueness proof of VIII is also simpler. In spite of these improvements, the construction of the reduced powers and proofs of properties constitute a lengthy and heavy piece of work. For this reason, we have adopted the axiomatic approach so that the reader will arrive quickly at the easier and more interesting parts.

The appendix, due to Epstein, presents purely algebraic proofs of propositions whose proofs, in the text, are mixed algebraic and geometric.

The reader should regard these lectures as an introduction to cohomology operations. There are a number of important topics which we have not included and which the reader might well study next. First, there is an alternate approach to cohomology operations based on the complexes $K(\pi,n)$ of Eilenberg-MacLane [Ann. of Math., 58 (1953), 55-106; 60 (1954), 49-139; 60 (1954), 513-555]. This approach has been developed extensively by H. Cartan [Seminar 1954/55]. A very important application of the squares has been made by J. F. Adams to the computation of the stable homotopy groups of spheres [Comment. Math. Helv. 32 (1958), 180-214]. Finally, we do not consider secondary cohomology operations. J. F. Adams has used these most successfully in settling the question of existence of mappings of spheres of Hopf invariant 1 [Ann. of Math., 72 (1960), 20-104; and Seminar, H. Cartan 1958/59].

N. E. S.

Princeton, New Jersey, May, 1962. D. B. A. E.

CONTENTS

COHOMOLOGY OPERATIONS

CHAPTER I.

Axiomatic Development of the Steenrod Algebra $\mathfrak{A}(2)$

In §1, axioms are given for Steenrod squares. (The existence and uniqueness theorems are postponed to the final chapters.) In §2, the effect of squares in projective spaces is discussed, and it is proved that any suspension of a Hopf map is essential. In §3, the algebra of the squares $\mathfrak{A}(2)$ is defined and the vector space basis of Adem [1] and Cartan [2] is obtained. In §4, it is shown that the indecomposable elements of the algebra $\mathfrak{A}(2)$ are represented by elements of the form Sq^{2^i}. Some geometric applications of this fact are given. In §5, the Hopf invariant of maps $S^{2n-1} \longrightarrow S^n$ is defined. The existence theorem for maps of even Hopf invariant when n is even, and some non-existence theorems, are given.

Unless otherwise stated, all homology and cohomology groups in this chapter will have coefficients Z_2.

§1. Axioms.

We now give axioms for the squares Sq^i. The existence and uniqueness theorems will be postponed to the final chapters.

1) For all integers $i \geq 0$ and $q \geq 0$, there is a natural transformation of functors which is a homomorphism

$$Sq^i\colon\ H^n(X,A) \longrightarrow H^{n+i}(X,A) \quad , \quad n \geq 0.$$

2) $Sq^0 = 1$.

3) If $\dim x = n$, $Sq^n x = x^2$.

4) If $i > \dim x$, $Sq^i x = 0$.

5) Cartan formula

$$Sq^k(xy) = \sum_{i=0}^{k} Sq^i x \,.\, Sq^{k-i} y \quad .$$

We recall that if $x \in H^p(X,A)$ and $y \in H^q(X,B)$, then $xy \in H^{p+q}(X,A \cup B)$. This is true in general in simplicial cohomology, but some condition of niceness on the subspaces A and B is necessary in singular cohomology.

6) Sq^1 is the Bockstein homomorphism β of the coefficient sequence

$$0 \longrightarrow Z_2 \longrightarrow Z_4 \longrightarrow Z_2 \longrightarrow 0 .$$

7) <u>Adem relations</u>. If $0 < a < 2b$, then

$$Sq^a Sq^b = \sum_{j=0}^{[a/2]} \binom{b-1-j}{a-2j} Sq^{a+b-j} Sq^j .$$

The binomial coefficient is, of course, taken mod 2.

The first five axioms imply the last two, as will be proved in the final chapter

1.1 LEMMA. The following two forms of the Cartan formula are equivalent in the presence of Axiom 1):

$$Sq^i(xy) = \Sigma_j \, Sq^j x \, . \, Sq^{i-j} y$$
$$Sq^i(x \times y) = \Sigma_j \, Sq^i x \times Sq^{i-j} y .$$

PROOF. Let $p\colon X \times Y \longrightarrow X$ and $q\colon X \times Y \longrightarrow Y$ be the projections. If the first formula holds, then

$$\begin{aligned} Sq^i(x \times y) &= Sq^i\big((x \times 1) \, . \, (1 \times y)\big) \\ &= \Sigma_j \, Sq^j(x \times 1) \, . \, Sq^{i-j}(1 \times y) \\ &= \Sigma_j \, Sq^j(p^* x) \, . \, Sq^{i-j}(q^* y) \\ &= \Sigma_j \, p^* Sq^i x \, . \, q^* Sq^{i-j} y \\ &= \Sigma_j \, (Sq^j x \times 1) \, . \, (1 \times Sq^{i-j} y) \\ &= \Sigma_j \, Sq^j x \times Sq^{i-j} y . \end{aligned}$$

Let $d\colon X \longrightarrow X \times X$ be the diagonal. If the second formula holds, then

$$\begin{aligned} Sq^i(xy) &= Sq^i d^*(x \times y) = d^* Sq^i(x \times y) \\ &= d^* \Sigma_j \, Sq^i x \times Sq^{i-j} y = \Sigma_j \, Sq^i x \, . \, Sq^{i-j} y . \end{aligned}$$

1.2 LEMMA. Axioms 1), 2) and 5) imply:

If $\delta\colon H^q(A) \longrightarrow H^{q+1}(X,A)$ is the coboundary map, then

$$\delta Sq^i = Sq^i \delta .$$

PROOF. We will show that δ is essentially equivalent to a $\times$-product with a 1-dimensional class. Then the Cartan formula applies to give the desired result. (This method can be used for any cohomology operation whose behaviour under $\times$-products is known.)

Let Y be the union of X and $I \times A$, with $A \subset X$ identified with $\{0\} \times A$. Let $B = [1/2,1]\times A \subset Y$ and $Z = X \cup [0,1/2]\times A \subset Y$, and let $A' = \{1\} \times A$ and $A'' = B \cap Z$. We then have the following commutative diagram.

$$\begin{array}{ccccccccc}
H^q(A) & \xrightarrow{\approx} & H^q(I \times A) & \xrightarrow{\approx} & H^q(A') & \xleftarrow{\text{epi}} & H^q(A' \cup Z) & \longrightarrow & H^q(A' \cup A'') \\
\downarrow \delta & & \downarrow \delta & & \downarrow \delta & & \downarrow \delta & & \downarrow \delta \\
H^{q+1}(X,A) & \xrightarrow{\approx} & H^{q+1}(Y,I\times A) & \xrightarrow{\approx} & H^{q+1}(Y,A') & \longleftarrow & H^{q+1}(Y,A' \cup Z) & \xrightarrow{\approx} & H^{q+1}(B,A' \cup A'')
\end{array}$$

The isomorphisms in the lower line are due to homotopy equivalence, the 5 lemma, and excision. In order to prove $\delta Sq^i = Sq^i\delta$ on $H^q(A)$, it is sufficient to prove it on $H^q(A' \cup Z)$. Looking at the last square on the right of the diagram, we see that it is sufficient to prove it on $H^q(A' \cup A'')$.

So we have to prove that $\delta Sq^i = Sq^i \delta$ where

$$\delta\colon\ H^q(\dot{I} \times A) \longrightarrow H^{q+1}(I \times A, \dot{I} \times A) .$$

Let $\bar{0}$ and $\bar{1}$ be the cohomology classes in $H^0(\dot{I})$ corresponding to the points 0 and 1. Let I be the generator of $H^1(I,\dot{I})$.

Starting with $\delta(\bar{1} \times u) = I \times u$, and applying the Cartan formula, we obtain

$$\begin{aligned} Sq^i\delta(\bar{1} \times u) &= Sq^i(I \times u) = Sq^0 I \times Sq^i u = I \times Sq^i u \\ &= \delta(\bar{1} \times Sq^i u) = \delta Sq^i(\bar{1} \times u) . \end{aligned}$$

Similarly, $\delta(\bar{0} \times u) = -I \times u$ leads to $Sq^i\delta(\bar{0} \times u) = \delta Sq^i(\bar{0} \times u)$.

§2. Projective Spaces.

Let $\underline{H}^q(X)$ denote the reduced cohomology group (mod 2).

2.1. LEMMA. Let SX denote the suspension of X, and let $s\colon\ \underline{H}^q(X) \longrightarrow \underline{H}^{q+1}(SX)$ denote the suspension isomorphism. Then, from Axioms 1), 2) and 5), it follows that $sSq^i = Sq^i s$.

PROOF. Let CX and $C'X$ be two cones on X. Then $SX = CX \cup C'X$, where $CX \cap C'X = X$. The suspension isomorphism is defined by the following commutative diagram of reduced cohomology groups

$$\begin{array}{ccc} \underline{H}^q(X) & \xrightarrow{\quad s \quad} & \underline{H}^{q+1}(SX) \\ \delta\big\downarrow\approx & & \approx\big\uparrow \\ \underline{H}^{q+1}(CX,X) & \xleftarrow[\approx]{\text{excision}} & \underline{H}^{q+1}(SX,C'X) \end{array}$$

The two vertical maps are isomorphisms because CX and $C'X$ are contractible. The lemma follows from Axiom 1) and 1.2.

2.2 LEMMA. If $X = \cup_{i=1}^{k} A_i$, where each A_i is open and contractible in X, then the product of any k positive-dimensional cohomology classes of X is zero.

PROOF. Since A_i is contractible in X, inclusion induces the zero homomorphism $H^q(X) \longrightarrow H^q(A_i)$ for $q > 0$. Hence $H^q(X,A_i) \longrightarrow H^q(X)$ is an epimorphism for each $q > 0$ and for each i. If u_i has positive dimension and $u_i \in H^*(X)$ for $1 \leq i \leq k$, then, for each i, there is an element $v_i \in H^*(X,A_i)$ which maps onto u_i. Now $v_1 v_2 \cdots v_k \in H^*(X, \cup A_i) = 0$ and the homomorphism $H^*(X,\cup A_i) \longrightarrow H^*(X)$ maps $v_1 v_2 \cdots v_k$ onto $u_1 u_2 \cdots u_k$. (Apply the theorem on the invariance of the cup-product under the inclusion $(X;\emptyset,\ldots,\emptyset) \subset (X;A_1,\ldots,A_k)$.) The lemma follows.

By 2.2, cup-products are zero in SX.

2.3. THEOREM. The n-fold suspension of the Hopf map $S^3 \longrightarrow S^2$ is essential.

PROOF. Let $X = P^2(C)$, the complex projective plane. One sees by Poincaré duality, that if x is the non-zero element of $H^2(X)$, then x^2 is the non-zero element of $H^4(X)$.

X is constructed by attaching the 4-cell e^4 to S^2 by means of the Hopf map $f\colon S^3 \longrightarrow S^2$. So S^nX is constructed by attaching the $(n+4)$-cell $S^n e^4$ to $S^{n+2} = S^n S^2$ by means of the map $S^n f\colon S^{n+3} \longrightarrow S^{n+2}$. Now

$$\begin{aligned} Sq^2(s^n x) &= s^n(Sq^2 x) && \text{by } 2.1 \\ &= s^n(x^2) && \text{by Axiom 3)} \\ &\neq 0 && \text{since } s \text{ is an isomorphism.} \end{aligned}$$

So $s^n x$ is the non-zero element of $H^{n+2}(S^nX)$ and $Sq^2(s^n x)$ is the non-zero element of $H^{n+4}(S^nX)$. Now suppose the map $S^n f$ is inessential. Then $S^nX \simeq S^{n+2} \vee S^{n+4}$. Let $r: S^nX \longrightarrow S^{n+2}$ be this homotopy equivalence followed by the obvious retraction. Let u be the non-zero element of $H^{n+2}(S^{n+2})$. Then $Sq^2u = 0$. So

$$0 = r^*(Sq^2u) = Sq^2(r^*u) = Sq^2(s^nx) \neq 0.$$

This is a contradiction.

We can prove in a similar manner that any suspension of the other Hopf maps is essential.

Axioms 3), 4) and 5) enable us to compute Sq^i on a part of the cohomology ring.

2.4. LEMMA. Axioms 2), 3), 4) and 5) imply that if $\dim u = 1$, then $Sq^iu^k = \binom{k}{i}u^{k+i}$.

PROOF. The lemma follows from Axioms 2) and 4) if $k = 0$. If $k > 0$, then by induction on k,

$$Sq^iu^k = Sq^i(u.u^{k-1}) = Sq^0u.Sq^iu^{k-1} + Sq^1u.Sq^{i-1}u^{k-1}$$

$$= \left[\binom{k-1}{i} + \binom{k-1}{i-1}\right]u^{k+1} = \binom{k}{i}u^{k+i} .$$

2.5. LEMMA. If $\dim u = 2$ and $Sq^1u = \beta u = 0$, then $Sq^{2i}(u^k) = \binom{k}{i}u^{k+i}$ and $Sq^{2i+1}(u^k) = 0$.

PROOF. This follows by induction, as in the previous lemma.

This following lemma is extremely useful in calculating binomial coefficients mod p.

2.6. LEMMA. Let p be a prime and let $a = \sum_{i=0}^m a_ip^i$ and $b = \sum_{i=0}^m b_ip^i$ $(0 \leq a_i, b_i < p)$. Then

$$\binom{b}{a} \equiv \prod_{i=0}^m \binom{b_i}{a_i} \mod p.$$

PROOF. $\binom{p}{i} = \dfrac{p(p-1)\ldots(p-i+1)}{1.2\ldots i} \quad (0 < i < p)$

$$\equiv 0 \mod p.$$

Therefore, in the polynomial ring $Z_p[x]$, we have $(1 + x)^p = 1 + x^p$. It follows by induction on i that $(1 + x)^{p^i} = 1 + x^{p^i}$. Therefore

$$(1+x)^b = (1+x)^{\Sigma b_i p^i} = \prod_{i=0}^{m} (1+x)^{b_i p^i}$$

$$= \prod_{i=0}^{m} (1+x^{p^i})^{b_i} = \prod_{i=0}^{m} \sum_{s=0}^{b_i} \binom{b_i}{s} x^{sp^i} .$$

The coefficient of $x^a = x^{\Sigma a_i p^i}$ in the usual expansion of $(1+x)^b$ is $\binom{b}{a}$. But, from the above expansion, we see that it is $\prod_{i=0}^{m} \binom{b_i}{a_i}$. The lemma follows.

2.7. LEMMA. If $\dim u = 1$, then

$$\begin{aligned} Sq^i\left(u^{2^k}\right) &= u^{2^k} && \text{if } i = 0 \\ &= 0 && \text{if } i \neq 0, 2^k \\ &= u^{2^{k+1}} && \text{if } i = 2^k . \end{aligned}$$

PROOF. This is immediate from 2.4 and 2.6.

§3. Definitions. The Basis of Admissible Monomials.

We now define the Steenrod algebra mod 2, $\mathfrak{A}(2)$. Let $M = (M_i)$ be a sequence of R-modules, where R is a commutative ring and $i \geq 0$. Then M is called a graded module. We say the elements of M_i have degree or dimension i. A homomorphism $f: A \longrightarrow B$ of graded modules is a sequence of homomorphisms $f_i: A_i \longrightarrow B_i$. If M and N are graded modules, we define the graded module $M \otimes N$ by $(M \otimes N)_r = \Sigma_i \; M_i \otimes N_{r-i}$. A graded R-module A is called a graded algebra if there is a homomorphism $\varphi: A \otimes A \longrightarrow A$ and a unit element 1 (which is obviously of degree 0). The algebra is said to be associative if commutativity holds in the diagram.

$$\begin{array}{ccc} A \otimes A \otimes A & \xrightarrow{\varphi \otimes 1} & A \otimes A \\ {\scriptstyle 1 \otimes \varphi}\downarrow & & \downarrow{\scriptstyle \varphi} \\ A \otimes A & \xrightarrow{\quad\varphi\quad} & A \end{array}$$

Let B be a graded module. Let $T: A \otimes B \longrightarrow B \otimes A$ be the map defined by $T(a \otimes b) = (-1)^{pq}(b \otimes a)$, where $p = \dim a$ and $q = \dim b$. We say the algebra is commutative if the diagram

$$\begin{array}{ccc} A \otimes A & \xrightarrow{T} & A \otimes A \\ & {\scriptstyle \varphi}\searrow \quad \swarrow{\scriptstyle \varphi} & \\ & A & \end{array}$$

is commutative. A homomorphism $f: A \longrightarrow B$ of algebras, is a homomorphism of modules, which commutes with the multiplication, i.e., $f\varphi_A = \varphi_B(f \otimes f)$, and such that $f(1) = 1$. Let M be a graded module and A a graded algebra. M is called an A-module, if there is a map $\psi: A \otimes M \longrightarrow M$, which respects the unit of A, and such that the following diagram is commutative

$$\begin{array}{ccc} A \otimes A \otimes M & \xrightarrow{1 \otimes \psi} & A \otimes M \\ \downarrow \varphi \otimes 1 & & \downarrow \psi \\ A \otimes M & \xrightarrow{\psi} & M \end{array}$$

If B is a graded algebra, then $A \otimes B$ is given a graded algebra structure by the multiplication $A \otimes B \otimes A \otimes B \xrightarrow{1 \otimes T \otimes 1} A \otimes A \otimes B \otimes B \xrightarrow{\varphi \otimes \varphi} A \otimes B$. If N is a B-module, then $M \otimes N$ is an $A \otimes B$-module by the mapping

$$A \otimes B \otimes M \otimes N \xrightarrow{1 \otimes T \otimes 1} A \otimes M \otimes B \otimes N \xrightarrow{\psi \otimes \psi} M \otimes N .$$

The ground ring R may be regarded as a graded module R, such that $R_i = 0$ if $i > 0$. We say a graded algebra is <u>augmented</u> if there is an algebra homomorphism $\varepsilon: A \longrightarrow R$. Let M be a graded R-module. Let M^r be the tensor product of M with itself r times, and let $\Gamma(M) = \sum_{n=0}^{\infty} M^n$ $(M^0 = R)$. $\Gamma(M)$ is called the tensor algebra of M. The multiplication $\Gamma(M) \otimes \Gamma(M) \longrightarrow \Gamma(M)$ is induced by the canonical isomorphisms $M^r \otimes M^s \approx M^{r+s}$.

We define $\mathfrak{a}(2)$, the Steenrod algebra mod 2, to be the graded associative algebra generated by the Sq^i, subject to the Adem relations (§ 1, Axiom 7). In detail, the construction is as follows. Let M be the graded Z_2-module, with $M_i \approx Z_2$ for all $i > 0$. We denote the generator of M_i by Sq^i, so that $\dim Sq^i = i$. $\mathfrak{a}(2)$ is the quotient of $\Gamma(M)$ by relations of the form

$$Sq^a \otimes Sq^b - \Sigma_j \binom{b-1-j}{a-2j} Sq^{a+b-j} \otimes Sq^j \quad \text{when } a < 2b .$$

We write $Sq^0 = 1$ in $\mathfrak{a}(2)$.

Given a sequence of non-negative integers $I = (i_1, i_2, \ldots, i_k)$, k is called the <u>length</u> of I. We write $k = \ell(I)$. We define the <u>moment</u> of I by $m(I) = \Sigma_{s=1}^{k} s i_s$. A sequence I is called <u>admissible</u> if both $i_{s-1} \geq 2i_s$ for $k \geq s \geq 2$, and $i_k \geq 1$. We write

$$Sq^I = Sq^{i_1} Sq^{i_2} \ldots Sq^{i_k}.$$

If I is admissible, we call Sq^I admissible. We also call Sq^0 admissible. We shall also speak of the moment of Sq^I.

3.1. THEOREM. The admissible monomials form a vector space basis for $\mathcal{A}(2)$.

PROOF. We first show that any inadmissible monomial is the sum of monomials of smaller moment, and hence that the admissible monomials span $\mathcal{A}(2)$. Let $I = (i_1, \ldots, i_k)$ be an inadmissible sequence with no zeros. For some r, $n = i_r < 2i_{r+1} = 2m$. So, by the Adem relations,

$$Sq^I = Sq^N Sq^n Sq^m Sq^M = \Sigma_j \lambda_j Sq^N Sq^{n+m-j} Sq^j Sq^M$$

where $\lambda_j \in Z_2$. It is easy to verify that each monomial on the right has smaller moment than Sq^I (separate arguments are needed for the cases $j = 0$ and $j > 0$). By induction on the moment, it follows that every monomial is a sum of admissible monomials.

We still need to show that the admissible monomials are linearly independent. Let P be ∞-dimensional real projective space. Then $H^*(P)$ is the polynomial ring $Z_2[u]$, where $\dim u = 1$. Let P^n be the n-fold Cartesian product of P with itself. Let $w = u \times u \ldots \times u \in H^n(P^n)$. The following proposition will complete the proof of 3.1.

3.2. PROPOSITION. The mapping $\mathcal{A}(2) \longrightarrow H^*(P^n)$, defined by evaluation on w, sends the admissible monomials of degree $\leq n$ into linearly independent elements.

PROOF. The proposition is proved by induction on n. For $n = 1$, it follows from 2.4.

Suppose $\Sigma a_I Sq^I w = 0$, where the sum is taken over admissible monomials Sq^I of a fixed degree q, where $q \leq n$. We wish to prove that $a_I = 0$ for each I. This is done by a decreasing induction on the length $\ell(I)$. Suppose that $a_I = 0$ for $\ell(I) > m$. The above relation takes the form

(1) $$\Sigma_{\ell(I)=m}\, a_I Sq^I w + \Sigma_{\ell(I)<m}\, a_I Sq^I w = 0 .$$

The Künneth theorem asserts that

$$H^{q+n}(P^n) \approx \Sigma_s H^s(P) \otimes H^{q+n-s}(P^{n-1}) .$$

Let g denote the projection into the summand with $s = 2^m$. Let $w = u \times w'$, where $w' \in H^{n-1}(P^{n-1})$. Then, by 1.1,

$$(2) \qquad Sq^I w = Sq^I(u \times w') = \Sigma_{J \leq I} Sq^J u \times Sq^{I-J} w' ,$$

where $J \leq I$ means $0 \leq j_r \leq i_r$ for all r. Let J_m be the sequence $(2^{m-1},\ldots,2^1,2^0)$. We assert that

$$(3) \qquad g\, Sq^I w = \begin{cases} 0 & \text{if } \ell(I) < m , \\ u^{2^m} \times Sq^{I-J_m} w' & \text{if } \ell(I) = m . \end{cases}$$

Recall that, by 2.7, $Sq^J u = 0$ unless J has the form $(2^{k-1}, 2^{k-2},\ldots, 2^1, 2^0)$ or is such a sequence interspersed with zeros. And $Sq^J u = u^{2^m}$ if $J = J_m$ or is obtained from J_m by interspersing zeros. In the last case $\ell(J) > m$. To prove 3), we refer to 2). If $\ell(I) < m$, then $J \leq I$ implies that $\ell(J) < m$, and so $g\, Sq^i w = 0$. If $\ell(I) = m$, then $g(Sq^J u \times Sq^{I-J} w') = 0$ unless $J = J_m \leq I$. This proves (3).

If we apply g to (1) and use (3), we find

$$(4) \qquad u^{2^m} \times \Sigma_{\ell(I)=m}\, a_I Sq^{I-J_m} w' = 0.$$

It is readily verified that, as I ranges over all admissible sequences of length m and degree q, $I - J_m$ will range over all admissible sequences of length $\leq m$ and degree $q - 2^m+1$; and the correspondence is one-to-one. Since $m \geq 1$, we have $q - 2^m + 1 \leq n - 1$. So the inductive hypothesis on n implies that each coefficient in (4) is zero. Thus $a_I = 0$ for $\ell(I) = m$.

This completes the proof of the proposition and hence of the theorem 3.1.

3.3. COROLLARY. The mapping $\mathfrak{a}(2) \longrightarrow H^*(P^n)$ given by $Sq^I \longrightarrow Sq^I w$ is a monomorphism in degrees $\leq n$.

EXERCISE. Find the basis of admissible monomials for $\mathfrak{a}_{12}$.

We note that, if I is an admissible sequence of length k, then $\deg\, Sq^I \geq 2^{k-1}+\ldots+1 = 2^k - 1$, so that the exercise is a finite problem.

§4. Indecomposable Elements.

Much of the material in this section is due to Adem [4].

Let A be an associative graded algebra. Let $\underline{A}$ be the ideal of A consisting of elements of positive degree. The set of decomposable elements of A is the image under $\varphi: A \otimes A \longrightarrow A$ of $\underline{A} \otimes \underline{A}$. This image is a two-sided ideal in A. $Q(A) = \underline{A}/\varphi(\underline{A} \otimes \underline{A})$ is called the set of indecomposable elements of A. A is called connected if $A_0 = R$, the ground ring.

4.1. LEMMA. In a graded connected algebra over a field, any set B of generators of A, contains a subset B_1, whose image in $Q(A)$ forms a vector space basis. Any such B_1 is minimal and generates A.

PROOF. Any set of generators of A spans $Q(A)$. Let B_1 be any subset of B whose image in $Q(A)$ is a basis. Let $g \in A$ be the element of smallest degree, which is not in the algebra A' generated by $\{1, B_1\}$. There is an element $g' \in A'$ such that $g - g'$ is decomposable. So $g - g' \in \varphi(\underline{A} \otimes \underline{A})$ and $g - g' = \Sigma a_i' a_i''$, where $a_i', a_i'' \in \underline{A}$. But a_i' and a_i'' are in A'. Therefore $g \in A'$, which is a contradiction.

4.2. LEMMA. Sq^i is decomposable if and only if i is not a power of 2.

PROOF. Writing the Adem relations in the form

$$\binom{b-1}{a} Sq^{a+b} = Sq^a Sq^b + \Sigma_{j>0} \binom{b-1-j}{a-2j} Sq^{a+b-j} Sq^j$$

where $0 < a < 2b$, one sees that if $\binom{b-1}{a} \equiv 1$, then Sq^{a+b} is decomposable. Suppose i is not a power of 2. Then $i = a + 2^k$, where $0 < a < 2^k$. Put $b = 2^k$. Then $b - 1 = 1 + \ldots + 2^{k-1}$. By 2.6 $\binom{b-1}{a} \equiv 1$. So, if i is not a power of 2, Sq^i is decomposable.

Now let $i = 2^k$. Suppose $Sq^{2^k} = \Sigma_{j=1}^{2^k-1} m_j Sq^j$. Then, using the notation of §2 and §3, we have by 2.7,

$$u^{2^{k+1}} = Sq^{2^k} u^{2^k} = \Sigma m_j \, Sq^j u^{2^k} = 0 .$$

This is a contradiction and the lemma is proved.

4.3. THEOREM. The elements Sq^{2^k} generate $\mathcal{A}(2)$ as an algebra

PROOF. This follows from 4.1 and 4.2.

We note that the elements Sq^{2^k} do not generate $\mathcal{A}(2)$ freely. In fact, by the Adem relations,

$$Sq^2Sq^2 = Sq^3Sq^1 = (Sq^1Sq^2)Sq^1 .$$

4.4. THEOREM. Let X be a space and let $x^2 \neq 0$, where $x \in H^q(X;Z_2)$. Then $Sq^{2^i}x \neq 0$ for some i such that $0 < 2^i \leq q$.

PROOF. $0 \neq x^2 = Sq^q x = \Sigma$ (monomials in Sq^{2^j})x where $2^j \leq q$ throughout the summation. The theorem follows.

A polynomial ring in one variable x is <u>truncated</u> if $x^n = 0$ for some $n \geq 2$.

4.5. THEOREM. If $H^*(X;Z_2)$ is a polynomial ring or a truncated polynomial ring on a generator x of dimension q, and $x^2 \neq 0$, then $q = 2^k$ for some k.

PROOF. Since $H^*(X)$ is a polynomial ring, $H^{q+2^i}(X) = 0$ for $0 < 2^i < q$. Therefore $Sq^{2^i}x = 0$ for $0 < 2^i < q$. By 4.4, $Sq^{2^k}x \neq 0$ for some k such that $0 < 2^k \leq q$. So $q = 2^k$.

REMARKS. J. F. Adams has shown [3] that the only possible values for k are 0,1,2,3. His methods entail a much deeper analysis of the algebra $\mathcal{A}(2)$.

Examples of spaces which satisfy the hypotheses of the theorem are

i) Real projective space of any dimension, with $q = 1$;

ii) Complex projective space of any dimension, with $q = 2$;

iii) Quaternionic projective space of any dimension, with $q = 4$;

iv) The Cayley projective plane with, $q = 8$.

4.6. THEOREM. Let M be a connected compact 2n-manifold, such that $H^q(M) = 0$ for $1 \leq q < n$, and with $H^n(M) = Z_2$. Then n is a power of 2.

PROOF. $H^{2n-q}(M) = 0$ for $1 \leq q < n$, and, if u is the generator of $H^n(M)$, u^2 is the generator of $H^{2n}(M)$. We now apply 4.5.

§5. The Hopf Invariant.

Let $f: S^{2n-1} \longrightarrow S^n$ $(n > 1)$. Let X be the adjunction space obtained by attaching a $2n$-cell e^{2n} to S^n by the mapping f. Then $H^n(X;Z) \approx Z$ and $H^{2n}(X;Z) \approx Z$, while for other positive dimensions the cohomology groups are zero. Let $x \in H^n(X;Z)$ and $y \in H^{2n}(X;Z)$ be generators. Then $x^2 = h(f).y$ for some integer $h(f)$ called the Hopf invariant of f. It is defined up to sign. A homotopy of f leaves the homotopy type of X unchanged, and so the Hopf invariant is an invariant of the homotopy class of f.

Sometimes the double covering $S^1 \longrightarrow S^1$ is assigned the Hopf invariant 1. In this case, the adjunction space is the projective plane.

5.1. THEOREM. If there exists a map $f: S^{2n-1} \longrightarrow S^n$ of odd Hopf invariant, then n is a power of 2.

PROOF. Let $\eta: H^*(X;Z) \longrightarrow H^*(X;Z_2)$ be the map induced by the coefficient homomorphism $Z \longrightarrow Z_2$. This map is a ring homomorphism. Hence $(\eta x)^2 = \eta y$, since $h(f) \equiv 1 \pmod 2$. By 4.5, n is a power of 2.

REMARKS. i) We easily see that, if n is odd, then $h(f) = 0$, for then $x^2 = -x^2$ and so $2x^2 = 0$ (integer coefficients).
2) The following are the standard maps of Hopf invariant one, and their adjunction space:

$S^3 \longrightarrow S^2$	complex projective plane;
$S^7 \longrightarrow S^4$	quaternionic projective plane;
$S^{15} \longrightarrow S^8$	Cayley projective plane.

5.2. THEOREM. (Hopf [5]). If n is even, there are maps $f: S^{2n-1} \longrightarrow S^n$ with any even Hopf invariant.

PROOF. Let S_1, S_2 and S be $(n-1)$-spheres and $f: S_1 \times S_2 \longrightarrow S$. We say f has degree (α,β) if $f|S_1 \times p_2$ has degree α and $f|p_1 \times S_2$ has degree β, where $(p_1,p_2) \in S_1 \times S_2$. The degree of f is independent of the choice of (p_1,p_2).

Let E_i be the n-cell such that $S_i = \text{bd}\, E_i$ $(i = 1,2)$. Now $\text{bd}(E_1 \times E_2) = (E_1 \times S_2) \cup (S_1 \times E_2)$ is a $(2n-1)$-sphere and

$(E_1 \times S_2) \cap (S_1 \times E_2) = S_1 \times S_2$. Let S' be the suspension of S. Then $S' = E_+ \cup E_-$ where E_+ and E_- are n-cells and $E_+ \cap E_- = S$.

Given a mapping $f\colon S_1 \times S_2 \longrightarrow S$, we extend f to a mapping

$$C(f)\colon (E_1 \times S_2) \cup (S_1 \times E_2) \longrightarrow E_+ \cup E_- = S'$$

in such a way that $C(f)(E_1 \times S_2) \subset E_+$ and $C(f)(S_1 \times E_2) \subset E_-$. $C(f)$ is a map $S^{2n-1} \longrightarrow S^n$.

5.2 will follow from two lemmas.

5.3. LEMMA. $h(C(f)) = \alpha\beta$.

PROOF. Throughout this proof integral coefficients will be used. Let X be the adjunction space $(E_1 \times E_2) \cup_{C(f)} S'$. The attaching map $C(f)$ gives rise to a map $g\colon (E_1 \times E_2, E_1 \times S_2, S_1 \times E_2) \longrightarrow (X, E_+, E_-)$. Let x be a generator of $H^n(X;Z)$. We define x_+ and x_- to be the inverse images of x under the isomorphisms $H^n(X,E_-) \longrightarrow H^n(X)$ and $H^n(X,E_+) \longrightarrow H^n(X)$ respectively. Now we have a map $(X,\emptyset,\emptyset) \longrightarrow (X,E_+,E_-)$. This gives rise to a commutative diagram

$$\begin{array}{ccc} H^n(X) \otimes H^n(X) & \longrightarrow & H^{2n}(X) \\ \uparrow & & \uparrow \\ H^n(X,E_+) \otimes H^n(X,E_-) & \longrightarrow & H^{2n}(X,S') \end{array} .$$

The vertical maps are isomorphisms. Therefore the cup-product $x_+ \cup x_-$ has image x^2 under the map $H^{2n}(X,S') \longrightarrow H^{2n}(X)$. We have the following commutative diagram

$$\begin{array}{ccccccc} H^n(X) & \xleftarrow{\approx} & & & H^n(X,E_-) & \xrightarrow{g^*} & H^n(E_1\times E_2, S_1\times E_2) \\ \downarrow{\scriptstyle\approx} & & & & \downarrow & & \downarrow{\scriptstyle\approx} \\ H^n(S') & \xleftarrow{\approx} & H^n(S',E_-) & \xrightarrow{\approx} & H^n(E_+,S) & \xrightarrow{g^*} & H^n(E_1\times p_2, S_1\times p_2) \\ & & & & {\scriptstyle\delta}\uparrow{\scriptstyle\approx} & & {\scriptstyle\delta}\uparrow{\scriptstyle\approx} \\ & & & & H^{n-1}(S) & \xrightarrow{g^*} & H^{n-1}(S_1\times p_2) \\ & & & & \uparrow{\scriptstyle\approx} & & \uparrow{\scriptstyle\approx} \\ & & & & Z & \xrightarrow{\alpha} & Z \end{array}$$

By the diagram $g^*x_+ = \alpha w_+$, where w_+ generates $H^n(E_1\times E_2, S_1\times E_2)$. By a similar diagram, we see that $g^*x_- = \beta w_-$, where w_- generates $H^n(E_1\times E_2, E_1\times S_2)$.

Let $p_i\colon E_1 \times E_2 \longrightarrow E_i$ $(i = 1,2)$. We define the generators $x_i \in H^n(E_i,S_i)$ by $p_1^* x_1 = w_+$ and $p_2^* x_2 = w_-$. Now

$$w_+ \cup w_- = p_1^* x_1 \cup p_2^* x_2 = (x_1 \times 1) \cup (1 \times x_2) = (x_1 \times x_2) .$$

Hence $g^*x_+ \cup g^*x_- = \alpha\beta(x_1 \times x_2)$ and $(x_1 \times x_2)$ generates $H^{2n}(E_1 \times E_2, E_1 \times S_2 \cup S_1 \times E_2)$.

Now $g\colon (E_1 \times E_2, E_1 \times S_2 \cup S_1 \times E_2) \longrightarrow (X,S')$ is a relative homeomorphism and therefore induces an isomorphism of cohomology groups. So we have the isomorphisms

$$H^{2n}(X) \xleftarrow{\approx} H^{2n}(X,S') \xrightarrow[g^*]{\approx} H^{2n}(E_1 \times E_2, E_1 \times S_2 \cup S_1 \times E_2) .$$

Under these isomorphisms $x^2 \in H^{2n}(X)$ corresponds to $x_+ \cup x_- \in H^{2n}(X,S')$ and to $\alpha\beta(x_1 \times x_2)$. Let y be the generator of $H^{2n}(X)$ which corresponds to $x_1 \times x_2$. Then $x^2 = \alpha\beta y$.

This proves the lemma.

5.4. LEMMA. There is a mapping $f\colon S^{n-1} \times S^{n-1} \longrightarrow S^{n-1}$ of type $(2,-1)$, if n is even.

PROOF. If $x,y \in S^{n-1}$, let $D(x)$ be the equatorial plane in Euclidean n-space R^n, having x as a pole. Let $f(x,y)$ be the image of y under the reflection through $D(x)$. If we represent x and y by vectors $(x_1,\dots,x_n)$ and $(y_1,\dots,y_n)$ in R^n, the mapping f is given by

$$f(x,y) = y - (2\textstyle\sum_{i=1}^n x_i y_i)\, x.$$

If we fix $x = (1,0,\dots,0)$, then $f(x,y) = (-y_1,y_2,\dots,y_n)$. This map has degree -1. If we fix $y = (1,0,\dots,0)$, then

$$f(x,y) = (1 - 2x_1^2, -2x_1x_2, \dots, -2x_1x_n) = g(x) .$$

g maps the plane $x_1 = 0$ into a point. It is one-to-one for $x_1 > 0$ and for $x_1 < 0$. g can be factored into $S^{n-1} \longrightarrow P^{n-1} \longrightarrow S^{n-1}$. The first map has degree 2 since n is even. The second has degree 1. Therefore g has degree 2 and the lemma is proved.

We can now complete the proof of 5.2. Let $f_1\colon S^{n-1} \longrightarrow S^{n-1}$ be any map of degree λ and $f_2\colon S^{n-1} \longrightarrow S^{n-1}$ be any map of degree μ. Then $g = f.(f_1 \times f_2)$ has degree $(2\lambda,-\mu)$, where f is the map of 5.4. By 5.3, the Hopf invariant of $C(g)$ is $-2\lambda\mu$.

REMARK. Suppose we have a real division algebra of finite dimension $n > 1$, with a two-sided unit and the multiplication map

$$m: R^n \times R^n \longrightarrow R^n.$$

Let S^{n-1} be the sphere with centre at 0, passing through the unit. Then we have a map

$$S^{n-1} \times S^{n-1} \xrightarrow{m} R^n - \{0\} \xrightarrow{r} S^{n-1} \quad (r = \text{radial projection from } 0)$$

which is of degree (1,1) since S^{n-1} contains the unit. By 5.3, we obtain a map of Hopf invariant one, $S^{2n-1} \longrightarrow S^n$. According to Adams [3], $n = 2, 4$ or 8.

BIBLIOGRAPHY

[1] J. Adem, "Relations on Steenrod Powers of Cohomology Classes," Algebraic Geometry and Topology, Princeton 1957.

[2] H. Cartan, "Sur l'itération des operations de Steenrod," Comment, Math. Helv., 29(1955), pp 40-58.

[3] J. F. Adams, "On the Non-Existence of Elements of Hopf Invariant One," Annals of Math., 72(1960), pp. 20-104.

[4] J. Adem, "The Iteration of the Steenrod Squares in Algebraic Topology," Proc. Nat. Acad. Sci. U.S.A., 38 (1952), pp. 720-726.

[5] H. Hopf, "Über die Abbildungen von Sphären auf Sphären niedrigerer Dimension," Fund. Math., 25(1935) pp. 427-440.

CHAPTER II.

The Dual of the Algebra $\mathfrak{A}(2)$

In §1 it is proved that the Steenrod algebra $\mathfrak{A}(2)$ is a Hopf algebra. The structure of the dual Hopf algebra is obtained in §2. In §3 it is proved that the algebra $\mathfrak{A}(2)$ is nilpotent. In §4 the canonical anti-automorphism c of a Hopf algebra is briefly discussed. In §5 various constructions with modules over the algebra $\mathfrak{A}(2)$ are described.

§1. The Algebra $\mathfrak{A}(2)$ is a Hopf Algebra.

1.1. THEOREM. The map of generators

$$\psi(Sq^k) = \Sigma_{i=0}^{k} Sq^i \otimes Sq^{k-i}$$

extends to a homomorphism of algebras $\psi\colon \mathfrak{A}(2) \longrightarrow \mathfrak{A}(2) \otimes \mathfrak{A}(2)$.

PROOF. Let $\underline{\mathfrak{A}}$ be the free associative algebra generated by the Sq^i $(i > 0)$. We have an epimorphism $\omega\colon \underline{\mathfrak{A}} \longrightarrow \mathfrak{A}$ (writing $\mathfrak{A}(2) = \mathfrak{A}$), with kernel generated by the Adem relations. The map ψ of generators extends naturally to an algebra homomorphism $\underline{\psi}\colon \underline{\mathfrak{A}} \longrightarrow \mathfrak{A} \otimes \mathfrak{A}$. We have to show that $\underline{\psi}$ vanishes on $\ker \omega$.

We have a map of modules

$$\alpha\colon H^*(X) \otimes H^*(Y) \longrightarrow H^*(X \times Y)$$

given by $\alpha(u \otimes v) = u \times v$. By the Künneth relations for a field, this is an isomorphism. Let P be ∞-dimensional real projective space. Let $X = P^n = P \times \ldots \times P$. Then, using the notation of Chapter I §3, the evaluation map on w, $w\colon \mathfrak{A} \longrightarrow H^*(X)$, is a monomorphism in degrees $\leq n$ by I 3.3. Therefore the map $w \otimes w\colon \mathfrak{A} \otimes \mathfrak{A} \longrightarrow H^*(X) \otimes H^*(X)$ is a monomorphism in degrees $\leq n$.

We have the diagram

$$\begin{array}{ccccc}
\mathfrak{A}\otimes\mathfrak{A} & \xrightarrow{w\otimes w} & H^*(X)\otimes H^*(X) & \underset{\approx}{\xrightarrow{\alpha}} & H^*(X\times X) \\
\uparrow{\scriptstyle \underline{\psi}} & & & & \uparrow{\scriptstyle w\times w} \\
\underline{\mathfrak{A}} & & \xrightarrow{\qquad\quad \omega \qquad\quad} & & \mathfrak{A}
\end{array}$$

We now prove that this diagram is commutative. $H^*(X)\otimes H^*(X)$ is an $\mathfrak{A}\otimes\mathfrak{A}$ module and hence, using the map $\underline{\psi}$, is an $\underline{\mathfrak{A}}$-module. Using the isomorphism α, this gives $H^*(X\times X)$ the structure of an $\underline{\mathfrak{A}}$-module. However, $H^*(X\times X)$ has its usual structure as an $\underline{\mathfrak{A}}$-module via ω. These two $\underline{\mathfrak{A}}$-modules are identical, for

$$\begin{aligned}
(\omega Sq^k)(u\times v) &= \Sigma\, Sq^i u\times Sq^{k-i}v \\
&= \alpha((\Sigma\, Sq^i\otimes Sq^{k-i})(u\otimes v)) \\
&= \alpha(\underline{\psi} Sq^k\,.\,u\otimes v).
\end{aligned}$$

Since the two $\underline{\mathfrak{A}}$-modules are identical, the diagram above is commutative.

Now, if $m\in\underline{\mathfrak{A}}$, $\deg m\leq n$, and $\omega m=0$, then, since the diagram is commutative and $w\otimes w$ is a monomorphism in dimensions $\leq n$, $\underline{\psi} m=0$. This completes the proof of the theorem.

Let A be an augmented graded algebra over a commutative ring R with a unit. We say A is a <u>Hopf algebra</u> if:

1) There is a "diagonal map" of algebras

$$\psi\colon\ A\longrightarrow A\otimes A;$$

2) The compositions

$$\begin{array}{ccccc}
 & & & A\otimes R & \\
 & & {\scriptstyle 1\otimes\varepsilon}\nearrow & & \searrow{\scriptstyle\approx} \\
A & \xrightarrow{\psi} & A\otimes A & & A \\
 & & {\scriptstyle \varepsilon\otimes 1}\searrow & & \nearrow{\scriptstyle\approx} \\
 & & & R\otimes A &
\end{array}$$

are both the identity.

We say ψ is <u>associative</u> if the diagram

$$\begin{array}{ccc}
A & \xrightarrow{\quad\psi\quad} & A\otimes A \\
\downarrow{\scriptstyle\psi} & & \downarrow{\scriptstyle\psi\otimes 1} \\
A\otimes A & \xrightarrow{1\otimes\psi} & A\otimes A\otimes A
\end{array}$$

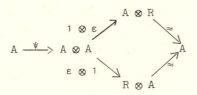

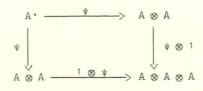

is commutative. We say ψ is <u>commutative</u> if the diagram

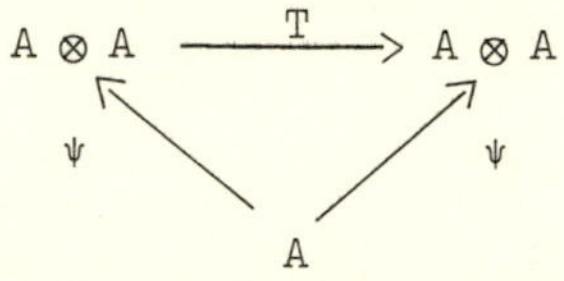

is commutative. (See I §3 for the definition of T.)

1.2. THEOREM. $\mathfrak{A}(2)$ is a Hopf algebra, with the commutative and associative diagonal map ψ of 1.1.

PROOF. The map ψ is a map of algebras by 1.1. Since $\mathfrak{A}(2)$ is connected, we have the unique augmentation ε: $\mathfrak{A}(2) \longrightarrow Z_2$. In the diagram

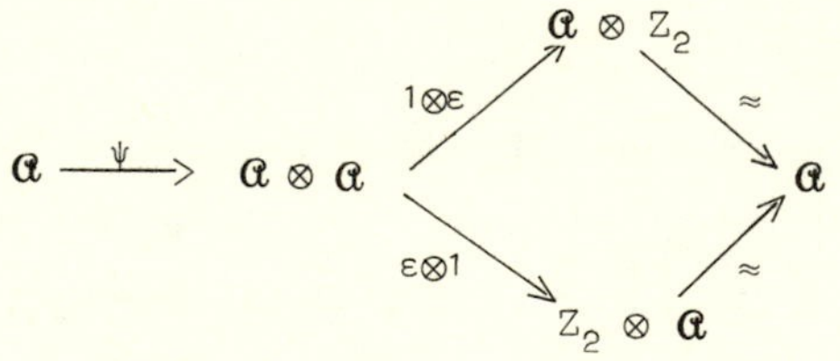

all the maps are homomorphisms of algebras. The compositions are both the identity on the generators of $\mathfrak{A}$, and they are therefore the identity on all of $\mathfrak{A}$. Using the fact that ψ is an algebra homomorphism, we see that ψ is commutative and associative by checking on the generators. This completes the proof.

Let A be a Hopf algebra with diagonal map ψ: $A \longrightarrow A \otimes A$. Let M be an A-module. Then $M \otimes M$ is an $A \otimes A$ -module. The map ψ defines an A-module structure on $M \otimes M$. Let m: $M \otimes M \longrightarrow M$ be a multiplication in M. We say that M is <u>an algebra over the Hopf algebra</u> A, if m is a homomorphism of A-modules.

1.3. PROPOSITION. If X is any space, $H^*(X;Z_2)$ is an algebra over the Hopf algebra $\mathfrak{A}(2)$.

PROOF. This results immediately from the Cartan formula, since ψ is a homomorphism of algebras.

Let X be a graded A-module, where A is a Hopf algebra over a ground ring R, with an associative diagonal map ψ. Let $\Gamma(X)$ be the tensor algebra of X over R. It is obvious that the usual multiplication $m: \Gamma(X) \otimes \Gamma(X) \longrightarrow \Gamma(X)$ is an A-homomorphism. Therefore $\Gamma(X)$ is an algebra over the Hopf algebra A.

§2. The Structure of the Dual Algebra

If X is a graded module over a field R, we say that X is of <u>finite type</u> if X_n is finite dimensional for each n. We define the dual X^* of X, to be the graded module with $X_n^* = \mathrm{Hom}(X_n,R)$. If X and Y are of finite type, then we have a canonical isomorphism $(X \otimes Y)^* \approx X^* \otimes Y^*$ defined by $(f \otimes g)(a \otimes b) = (-1)^{pq} fa \otimes gb$, $p = \deg a$, $q = \deg g$.

If A is a Hopf algebra of finite type, with multiplication φ and diagonal ψ, we easily verify that A^* is also a Hopf algebra, with multiplication ψ^* and diagonal φ^*.

For $k > 0$, let $M_k = Sq^I$, where $I = (2^{k-1}, 2^{k-2}, \ldots, 2, 1)$. M_k is an admissible monomial in $\mathfrak{A}$. Let $\xi_k \in \mathfrak{A}^*$ be the dual of M_k with respect to the basis of admissible monomials in $\mathfrak{A}$. Then $\langle \xi_k, M_k \rangle = 1$ and $\langle \xi_k, m \rangle = 0$ if m is admissible and $m = M_k$. M_k has degree $2^k - 1$ and therefore ξ_k has degree $2^k - 1$.

Let P be ∞-dimensional real projective space. Let $x \in H^1(P;Z_2)$ be the generator. Let $P^n = P \times P \ldots \times P$. In $H^n(P^n;Z_2)$ we have the element $x_1 \times x_2 \times \ldots \times x_n$, where each $x_i = x$. The following theorem, together with I 3.3, will enable us to find the structure of $\mathfrak{A}^*$.

By induction on n, we shall define $x(I) \in H^*(P^n)$ and $\xi(I) \in \mathfrak{A}^*$, where $I = (i_1, \ldots, i_n)$ is a sequence of non-negative integers. If $I = (i)$, we put $x(I) = x^{2^i}$ and $\xi(I) = \xi_i$. Suppose $x(I)$ and $\xi(I)$ are defined when I has length less than n. Now suppose $I = (i_1, \ldots, i_n)$. We put $x(I) = x(i_1) \times x(i_2, \ldots, i_n)$ and $\xi(I) = \xi(i_1)\, \xi(i_2, \ldots, i_n)$.

2.1. THEOREM. If $\alpha \in \mathfrak{A}$, then

$$\alpha(x_1 \times \ldots \times x_n) = \Sigma_{\ell(I)=n} \langle \xi(I), \alpha \rangle\, x(I) .$$

(The summation is finite, since $\langle \xi(I), \alpha \rangle = 0$ unless $\xi(I)$ and α have the same dimension.)

PROOF. We prove the theorem by induction on n. If α is admissible, then $\alpha x = 0$ unless $\alpha = M_k$, and $M_k x = x^{2^k}$ by I 2.7. The formula is therefore true for $n = 1$, when α is admissible. Since each element in $\mathfrak{A}$ is the sum of admissible monomials, the theorem is true for $n = 1$.

We now assume the theorem is true for integers less than n. Let $\psi\alpha = \Sigma_i \alpha_i' \otimes \alpha_i''$. By 1.3 and 1.1.

$$\begin{aligned}\alpha(x_1 \times \ldots \times x_n) &= \Sigma_i \alpha_i' x_1 \times \alpha_i''(x_2 \times \ldots \times x_n) \\ &= \Sigma_{i,I} < \xi(i_1),\alpha_i' > < \xi(i_2,\ldots,i_n),\alpha_i'' > x(I) \\ &= \Sigma_{i,I} < \xi(i_1) \otimes \xi(i_2,\ldots,i_n),\alpha_i' \otimes \alpha_i'' > x(I) \\ &= \Sigma_I < \xi(i_1) \otimes \xi(i_2,\ldots,i_n),\psi\alpha > x(I) \\ &= \Sigma_{\ell(I)=n} < \xi(I),\alpha > x(I).\end{aligned}$$

The last line follows since ψ^* is the multiplication in $\mathfrak{A}^*$. This completes the proof of the theorem.

We can now find the structure of $\mathfrak{A}^*$ as an algebra.

Let $\mathfrak{A}'$ be the polynomial algebra over Z_2, generated by the elements $\xi_1,\xi_2,\ldots$. Since ψ is commutative, the multiplication ψ^* in $\mathfrak{A}^*$ is commutative. So we have a homomorphism of algebras $\mathfrak{A}' \longrightarrow \mathfrak{A}^*$, defined in the obvious way.

2.2. THEOREM. (Milnor [1]). The map $\mathfrak{A}' \longrightarrow \mathfrak{A}^*$ is an isomorphism.

PROOF. We first show that $\mathfrak{A}' \longrightarrow \mathfrak{A}^*$ is an epimorphism. Suppose $< \xi(I),\alpha > = 0$ for all choices of I. By 2.1, we then have $\alpha(x_1 \times \ldots \times x_n) = 0$ for all n. But, by I 3.3, this shows that $\alpha = 0$. So the annihilator of $\mathrm{Im}(\mathfrak{A}' \longrightarrow \mathfrak{A}^*)$ is zero. Therefore $\mathfrak{A}' \longrightarrow \mathfrak{A}^*$ is an epimorphism.

We now show that the map $\mathfrak{A}' \longrightarrow \mathfrak{A}^*$ is an isomorphism by showing that in each dimension the ranks of $\mathfrak{A}'$ and $\mathfrak{A}^*$ as vector spaces over Z_2 are the same. We have only to show that the ranks of $\mathfrak{A}'$ and $\mathfrak{A}$ are the same in each dimension.

We write $\xi^I = \xi_1^{i_1}\xi_2^{i_2}\ldots\xi_n^{i_n}$, where $I = (i_1,i_2,\ldots,i_n,0,\ldots)$.

The monomials ξ^I in $\mathfrak{A}'$ thus correspond in a one-to-one way with sequences of non-negative integers $(i_1, i_2, \ldots, i_n, 0, \ldots)$. The admissible monomials $Sq^{I'} \in \mathfrak{A}$ correspond to sequences of integers $(i_1', i_2', \ldots, i_n', 0, \ldots)$ where $i_k' \geq 2i_{k+1}'$ and $i_n' \geq 1$. It remains only to set up a one-to-one correspondence between sequences of non-negative integers I and admissible sequences I' such that ξ^I and $Sq^{I'}$ have the same degree.

Let I_k be the sequence which is zero everywhere except for a 1 in the kth place. Let

$$I_k' = (2^{k-1}, 2^{k-2}, \ldots 2, 1, 0, 0, \ldots).$$

We construct a map from the set of sequences I to the set of sequences I' by insisting that I_k be sent to I_k' and that the map be additive (with respect to coordinatewise addition). Then if

$$I = (i_1, \ldots, i_n, 0, \ldots) \longrightarrow I' = (i_1', \ldots, i_n', 0, \ldots)$$

ξ^I and $Sq^{I'}$ have the same degree and we have

$$i_k' = i_k + 2i_{k+1} + \ldots + 2^{n-k} i_n.$$

Solving for i_k in terms of i_k', we obtain

$$i_k = i_k' - 2i_{k+1}' .$$

Therefore every admissible sequence I' is the image of a unique sequence I of non-negative integers. Thus the correspondence is one-to-one.

This completes the proof of the theorem.

We now find the diagonal in $\mathfrak{A}^*$.

2.3. THEOREM. (Milnor [1]). The diagonal map $\varphi^*: \mathfrak{A}^* \longrightarrow \mathfrak{A}^* \otimes \mathfrak{A}^*$ is given by

$$\varphi^* \xi_k = \Sigma_{i=0}^{k} \xi_{k-i}^{2^i} \otimes \xi_i .$$

PROOF. Let $\alpha, \beta \in \mathfrak{A}$. We have to show that

$$< \varphi^* \xi_k, \alpha \otimes \beta > = < \Sigma\, \xi_{k-i}^{2^i} \otimes \xi_i, \alpha \otimes \beta > .$$

That is, we have to show

$$< \xi_k, \alpha\beta > = \Sigma_i < \xi_{k-i}^{2^i}, \alpha > < \xi_i, \beta > .$$

We shall prove this by using 2.1.

Let x be the generator of $H^1(P; Z_2)$. Let $d\colon P \longrightarrow P^n$ be the diagonal, where $n = 2^i$. Then

$$x^{2^i} = d^*(x_1 \times \ldots \times x_n) .$$

So

$$\begin{aligned} \alpha \cdot x^{2^i} &= \alpha \cdot d^*(x_1 \times \ldots \times x_n) . \\ &= d^* \alpha(x_1 \times \ldots \times x_n) \\ &= d^*(\Sigma_{\ell(I)=n} < \xi(I), \alpha > x(I)) \\ &= \Sigma_{\ell(I)=n} < \xi(I), \alpha > x^{n(I)} \end{aligned}$$

where $n(I) = 2^{i_1} + \ldots + 2^{i_n}$ if $I = (i_1, \ldots, i_n)$. If we cyclically permute I, we dc not alter $< \xi(I), \alpha > x^{n(I)}$. Since $n = 2^i$, the number of different sequences, obtainable by cyclic permutation from one particular sequence I, is some power of 2, say $2^{j(I)}$. If $j(I) > 0$, the terms in the summation corresponding to cyclic permutations of I will cancel out mod 2. So we are left with terms for which $j(I) = 0$. That is, $m = i_1 = i_2 = \ldots = i_n$. For such sequences I, $\xi(I) = \xi_m^{2^i}$ and $n(I) = 2^{m+i}$. Therefore

$$\alpha x^{2^i} = \Sigma_m < \xi_m^{2^i}, \alpha > x^{2^{m+i}} .$$

Now

$$\begin{aligned} \Sigma_k < \xi_k, \alpha\beta > x^{2^k} &= \alpha\beta \cdot x \qquad \text{by } 2.1 \\ &= \alpha \cdot \beta x \\ &= \alpha \, \Sigma_i < \xi_i, \beta > x^{2^i} \\ &= \Sigma_{i,m} < \xi_m^{2^i}, \alpha > < \xi_i, \beta > x^{2^{m+i}} . \end{aligned}$$

Equating coefficients of x^{2^k}, we see that

$$< \xi_k, \alpha\beta > = \Sigma_{m+i=k} < \xi_m^{2^i}, \alpha > < \xi_i, \beta >$$

which proves our theorem.

§3. Ideals

Let A be a Hopf algebra of finite type over a field, with diagonal ψ. An ideal M is called a Hopf ideal, if $\psi(M) \subset M \otimes A + A \otimes M$. If M is a Hopf ideal, then A/M has an induced Hopf algebra structure (assuming $1 \notin M$). If A^* is the dual of A and $M^\dagger$ the dual of A/M, then $M^\dagger$ is the Hopf subalgebra of A^* which annihilates M. Conversely, if $M^\dagger$ is a Hopf subalgebra of A^*, then the dual algebra to $M^\dagger$ is the quotient of A by the Hopf ideal M which annihilates $M^{\dagger}$

In the algebra $\mathfrak{A}(2)^*$, let $M(j_1,\ldots,j_k,\ldots)$ be the ideal generated by the elements ξ_k^n, where $n = 2^{j_k}$ $(k = 1,2,\ldots)$.

3.1. LEMMA. If $j_{k-1} \leq j_k + 1$ for all k, then M is a Hopf ideal.

PROOF. $\varphi^*\xi_k^n = (\varphi^*\xi_k)^n = \Sigma_{i=0}^k \xi_{k-i}^{2^i.n} \otimes \xi_i^n$ if n is a power of 2. By induction on i, $j_{k-i} \leq j_k + i$. Therefore, if $i < k$, $\xi_{k-i}^{2^i.n} \in M$, where $n = 2^{j_k}$. In the term of the summation where $i = k$, we have $\xi_k^n \in M$ where $n = 2^{j_k}$. This proves 3.1.

Let M_h be the ideal of the sequence $(h,h-1,\ldots,1,0,0\ldots)$. Let $\mathfrak{A}_h$ be the Hopf subalgebra of $\mathfrak{A}$ which annihilates M_h. Since $\mathfrak{A}^*/M_h$ is finite, so is $\mathfrak{A}_h$.

3.2. LEMMA. $Sq^i \in \mathfrak{A}_h$ for $i < 2^h$.

PROOF. The proof is by induction on i. It is obvious for $i = 0$. We must show that $\xi_k^r\xi^J . Sq^i = 0$ if $r = \max(1, 2^{h-k+1})$ and J is arbitrary. Now

$$\begin{aligned} \varphi^*(\xi_k^r \otimes \xi^J) . Sq^i &= (\xi_k^r \otimes \xi^J) . \psi Sq^i \\ &= \Sigma_j (\xi_k^r Sq^j)(\xi^J Sq^{i-j}) \\ &= \xi_k^r Sq^i . \xi^J Sq^0 \end{aligned}$$

by our induction hypothesis. Now

$$\deg \xi_k^r = r(2^k - 1) \geq 2^{h-k+1}(2^k - 1) = 2^{h+1} - 2^{h-k+1} \geq 2^h.$$

Also $\deg Sq^i = i < 2^h$. Therefore $\xi_k^r Sq^i = 0$. This completes the proof of the lemma.

REMARK. Actually the elements Sq^{2^i} $(i < h)$ generate $\mathfrak{A}_h$, but we shall not prove this.

3.3. COROLLARY. $\mathfrak{A}$ is the union of the sequence $\mathfrak{A}_h$ $(h=1,2,\ldots)$ each a finite Hopf subalgebra of $\mathfrak{A}$.

3.4. LEMMA. If $\xi \in \mathfrak{A}^*$, then

$$\begin{aligned} \xi^2 \, . \, Sq^I &= \xi \, . \, Sq^J \quad \text{if } I = 2J \\ &= 0 \qquad\quad \text{otherwise.} \end{aligned}$$

PROOF.
$$\begin{aligned} \xi^2 \, . \, Sq^I &= \psi^*(\xi \otimes \xi) \, . \, Sq^I \\ &= (\xi \otimes \xi) \, . \, \psi Sq^I \\ &= \xi \otimes \xi \, . \, \Sigma_{R+S=I}\, Sq^R \otimes Sq^S \\ &= \Sigma_{R+S=I}\, (\xi Sq^R)(\xi Sq^S) \, . \end{aligned}$$

If we interchange R and S, we do not alter $(\xi Sq^R)(\xi Sq^S)$. Therefore the terms of the summation cancel mod 2, unless $R = S = J$, when $I = 2J$. Now if $x \in Z_2$, then $x^2 = x$. Therefore $(\xi Sq^J)^2 = \xi Sq^J$. The lemma follows.

If A is any commutative algebra over Z_2 and $\lambda\colon A \longrightarrow A$ is defined by $\lambda x = x^2$, then λ is a map of algebras. Moreover, λ commutes with maps of algebras. Hence if A is a Hopf algebra, λ is a map of Hopf algebras.

Then $\lambda\colon \mathfrak{A}^* \longrightarrow \mathfrak{A}^*$ doubles degrees. λ is a monomorphism, since the elements ξ^{2I} as I varies, are linearly independent.

Let $\lambda^*\colon \mathfrak{A} \longrightarrow \mathfrak{A}$ be the dual map. Then λ^* is an epimorphism of Hopf algebras. Since λ doubles degrees and misses odd degrees, λ^* divides even degrees by two and sends elements of odd degree to zero.

3.5. PROPOSITION.
$$\begin{aligned} \lambda^*(Sq^I) &= Sq^J \quad \text{if } I = 2J \\ &= 0 \qquad \text{otherwise.} \end{aligned}$$

The kernel of λ^* is the ideal generated by Sq^1.

PROOF.
$$\begin{aligned} \xi \, . \, (\lambda^* Sq^I) &= \lambda\xi \, . \, Sq^I \\ &= \xi^2 \, . \, Sq^I = \xi \, . \, Sq^J \qquad \text{if } I = 2J \\ &\qquad\qquad\;\; = 0 \, . \qquad\quad \text{otherwise.} \end{aligned}$$

This proves the first part of the proposition.

Let $m = Sq^{I_1} + \ldots + Sq^{I_n}$ be a sum of admissible monomials. Then, if $I_r = 2J_r$ for some r, $\lambda^* m$ is a sum of admissible monomials containing the term Sq^{J_r}. So, if $\lambda^* m = 0$, I_r is not divisible by 2 for any r; that is, Sq^{I_r} has a factor Sq^{2i+1}. Now we have the Adem relation

$$Sq^1 Sq^{2i} = \binom{2i-1}{1} Sq^{2i+1} = Sq^{2i+1} \, .$$

So, $Sq^{2i+1} \in \{Sq^1\}$ and therefore $m \in \{Sq^1\}$ if $\lambda^* m = 0$. So, $\ker \lambda^* \subset \{Sq^1\}$. On the other hand, since $\lambda^* Sq^1 = 0$, we also have $\{Sq^1\} \subset \ker \lambda^*$. This completes the proof of the proposition.

3.6. COROLLARY. If S_h is the ideal of $\mathcal{A}$ generated by Sq^n for $n = 2^0, 2^1, \ldots, 2^{h-1}$, then $(\lambda^*)^h\colon \mathcal{A} \longrightarrow \mathcal{A}$ has kernel S_h, and so S_h is a Hopf ideal. The map $(\lambda^*)^h$ is given as follows:

$$\begin{aligned} Sq^I &\longrightarrow Sq^J && \text{if } I = 2^h J \\ &\longrightarrow 0 && \text{otherwise.} \end{aligned}$$

This map induces an isomorphism of Hopf algebras $\mathcal{A}/S_h \longrightarrow \mathcal{A}$.

PROOF. This follows by induction on h.

EXERCISE. Let $[\mathcal{A}, \mathcal{A}]$ be the ideal of $\mathcal{A}$ generated by all the commutators $\alpha\beta - \beta\alpha$ $(\alpha, \beta \in \mathcal{A})$. $[\mathcal{A}, \mathcal{A}]$ is a Hopf ideal and $\mathcal{A}/[\mathcal{A}, \mathcal{A}]$ is a divided polynomial algebra on one generator; i.e.,

$$Sq^i Sq^j = \binom{i+j}{j} Sq^{i+j} .$$

(Hint: Prove the dual proposition in $\mathcal{A}^*$.)

§4. The Conjugation c

Let A be a connected Hopf algebra over a field with associative diagonal ψ and multiplication φ. We define a map $c\colon A \longrightarrow A$ by induction on dimension. Let $c(1) = 1$. If $\psi x = x \otimes 1 + \sum x_i' \otimes x_i'' + 1 \otimes x$, we define $cx = -x - \sum_i (cx_i')x_i''$. Let A' be the opposite Hopf algebra. That is, $A' = A$ as a graded vector space, and the multiplication φ' and diagonal ψ' are defined by commutativity of the diagram

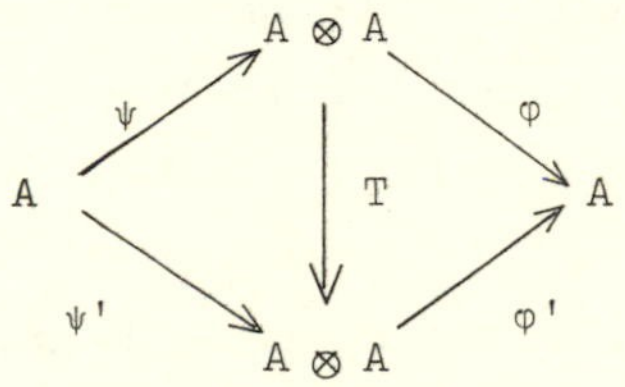

For the proof of the following theorem we refer the reader to the final chapter of "On the Structure of Hopf Algebras," by Moore and Milnor, to appear in Transactions of the A. M. S.

4.1. THEOREM. The map $c\colon A \longrightarrow A'$ is an isomorphism of Hopf algebras. If A has either a commutative diagonal or a commutative multiplication, then $c^2 = 1$.

The motivation for the definition of c is as follows. If G is a compact connected Lie group and K is a field, then $H_*(G;K)$ is a Hopf algebra over K with diagonal ψ induced by the diagonal $G \longrightarrow G \times G$ and the multiplication φ induced by the multiplication in G. The map c is induced by the map $g \longrightarrow g^{-1}$ of G. We easily see that $\varphi(c \otimes 1)\psi$ is induced by the map $g \longrightarrow 1$, and that the formula above for c is therefore satisfied. In this case 4.1 is obvious.

In $\mathfrak{A}$, we have

$$\begin{aligned} c(Sq^1) &= Sq^1; \\ c(Sq^2) &= Sq^2 + Sq^1Sq^1 = Sq^2; \\ c(Sq^3) &= Sq^3 + Sq^1Sq^2 + Sq^2Sq^1 = Sq^2Sq^1; \\ c(Sq^4) &= Sq^4 + Sq^1Sq^3 + Sq^2Sq^2 + Sq^2Sq^1Sq^1 \\ &= Sq^4 + Sq^3Sq^1; \end{aligned}$$

etc.

§5. Unstable $\mathfrak{A}$-modules

We define the excess of $Sq^I = Sq^{i_k} \dots Sq^{i_1}$ to be $(i_k - 2i_{k-1}) + (i_{k-1} - 2i_{k-2}) + \dots + (i_2 - 2i_1) + i_1$. The excess is non-negative for an admissible monomial. Let $x \in H^n(X)$. If $Sq^I x \neq 0$, then $i_k \leq n + i_1 + \dots + i_{k-1}$ by Axiom 4), I §1. We define $B(n)$ to be the subspace of $\mathfrak{A}$ spanned by all monomials Sq^I which can be factored into the form $m_1 Sq^i m_2$, where m_1 and m_2 are monomials and $i > n + \deg m_2$. It is obvious that $B(n)$ is a left ideal which annihilates all cohomology classes of dimensions $\leq n$. Any admissible monomial of excess greater than n is in $B(n)$, since the excess is $i_k - (i_{k-1} + \dots + i_1)$.

5.1. LEMMA. $B(n)$ is the vector space spanned by all admissible monomials of excess greater than n.

PROOF. We shall show that, on applying an Adem relation to a monomial in $B(n)$ we obtain a sum of monomials in $B(n)$. By repeated application of Adem relations, we then express the monomial as a sum of admissible monomials in $B(n)$. Any admissible monomial in $B(n)$ has excess greater than n and so the lemma will follow.

Suppose then that in the monomial $m_1 Sq^i m_2$, $i > n + \deg m_2$. Applying an Adem relation to either m_1 or m_2, we get a sum of monomials of the same form. If $i < 2b$, and $m_2 = Sq^b m_2'$, then

$$m_1 Sq^i Sq^b m_2' = \sum_{t=0}^{[i/2]} \binom{b-1-t}{i-2t} m_1 Sq^{i+b-t} Sq^t m_2' .$$

Now

$$i + b - t > i > n + \deg m_2 = n + \deg m_2' + b > n + \deg m_2' + t.$$

If $a < 2i$, and $m_1 = m_1' Sq^a$, then

$$m_1' Sq^a Sq^i m_2 = \sum_{t=0}^{[a/2]} \binom{i-1-t}{a-2t} m_1' Sq^{a+i-t} Sq^t m_2 .$$

Now

$$a + i - t > n + \deg m_2 + a - t \geq n + \deg m_2 + t = n + \deg(Sq^t m_2).$$

The lemma follows.

Suppose X is an $\mathcal{A}$-module. We say X is an <u>unstable $\mathcal{A}$-module</u>, if $B(n)X_n = 0$ for all $n \geq 0$. This is equivalent to the assertion $Sq^i x = 0$ if $i > \dim x$. The category of unstable $\mathcal{A}$-modules and $\mathcal{A}$-maps is a subcategory of the category of $\mathcal{A}$-modules and $\mathcal{A}$-maps. This category is closed if one takes:

1) Submodules
2) Quotient modules
3) Direct sums
4) Tensor products over Z_2.

Only the last needs proof. If X and Y are $\mathcal{A}$-modules, then $X \otimes Y$ is an $\mathcal{A}$-module through the diagonal map. So

$$Sq^i(x \otimes y) = \Sigma\, Sq^j x \otimes Sq^{i-j} y .$$

If $i > \dim x + \dim y$, then either $j > \dim x$ or $i-j > \dim y$, and so $Sq^i(x \otimes y) = 0$.

Let $F(n)$ be the $\mathcal{A}$-module defined by: $F(n)_i$ is the image of $\mathcal{A}_{i-n}$ in $\mathcal{A}/B(n)$. Then it is easy to see that $F(n)$ is an unstable $\mathcal{A}$-module. $F(n)$ is called the <u>free unstable $\mathcal{A}$-module on one n-dimensional</u>

generator. A free unstable $\mathfrak{A}$-module is the direct sum of free unstable $\mathfrak{A}$-modules on one generator.

5.2. PROPOSITION. Any unstable $\mathfrak{A}$-module is the quotient of a free unstable $\mathfrak{A}$-module.

PROOF. The proof is the same as the standard proof for modules.

5.3. LEMMA. Let X be an unstable $\mathfrak{A}$-module and $\Gamma(X)$ its tensor algebra (see end of §1). Let D be the ideal of $\Gamma(X)$ generated by all elements of the forms $x \otimes y - (-1)^{mn} y \otimes x$ and $Sq^n x - x \otimes x$ ($m = \dim y$, $n = \dim x$) for all $x, y \in X$. Then D is an $\mathfrak{A}$-ideal. Hence $\Gamma(X)/D$ is an $\mathfrak{A}$-algebra.

PROOF. If $i > 2k$ and $\dim x = k$, then

$$Sq^i(Sq^k x - x \otimes x) = Sq^i Sq^k x - \Sigma_j\ Sq^j x \otimes Sq^{i-j} x = 0.$$

If $i = 2k$,

$$Sq^i(Sq^k x - x \otimes x) = Sq^{2k} Sq^k x - Sq^k x \otimes Sq^k x = Sq^{2k} y - y \otimes y .$$

If $i < 2k$,

$$Sq^i(Sq^k x - x \otimes x) = \sum_{t=0}^{[i/2]} \binom{k-1-t}{i-2t} Sq^{i+k-t} Sq^t x - \Sigma_j\ Sq^j x \otimes Sq^{i-j} x.$$

Now $Sq^{i+k-t} Sq^t x = 0$ if $i + k - t > k + t$, ie., if $i > 2t$. Cancelling mod 2,

$$\Sigma\ Sq^i x \otimes Sq^{i-j} x = \begin{cases} 0 & \text{if } i \text{ is odd,} \\ Sq^{i/2} x \otimes Sq^{i/2} x & \text{if } i \text{ is even.} \end{cases}$$

So

$$Sq^i(Sq^k x - x \otimes x) = 0 \qquad \text{if } i \text{ is odd,}$$

$$\begin{aligned} Sq^i(Sq^k x - x \otimes x) &= Sq^{k+i/2} Sq^{i/2} x - Sq^{i/2} x \otimes Sq^{i/2} x \qquad \text{if } i \text{ is even,} \\ &= Sq^{k+i/2} y - y \otimes y. \end{aligned}$$

Also

$$Sq^i(x_1 \otimes x_2 - x_2 \otimes x_1) = \Sigma_j\ (Sq^j x_1 \otimes Sq^{i-j} x_2 - Sq^{i-j} x_2 \otimes Sq^j x_1) .$$

Finally, we must show that, if r is a relation, and $\alpha\beta \in \Gamma(X)$, then $Sq^i(\alpha\ r\ \beta)$ is in the ideal.

$$Sq^i(\alpha\ r\ \beta) = \Sigma_{h+s+t=i}\ Sq^h \alpha \,.\, Sq^s r \,.\, Sq^t \beta.$$

Since $Sq^s r$ is in the ideal, so is $Sq^i(\alpha\, r\, \beta)$.

5.4. DEFINITION. If $X, \Gamma(X)$ and D are as in 5.3, then the quotient algebra $\Gamma(X)/D$ is denoted by $U(X)$ and is called the <u>free $\mathfrak{A}$-algebra</u> generated by X. Let M be a free unstable $\mathfrak{A}$-module. Then $U(M)$ is called a <u>completely free $\mathfrak{A}$-algebra</u>.

Let $K(G,n)$ denote the Eilenberg-MacLane complex of the group G in dimension n. The cohomology $H^*(K(Z_2,n);Z_2)$ has been computed by J. P. Serre, Comment. Math. Helv. 27(1953), 198-232. His result can be restated: $H^*(K(Z_2,n);Z_2)$ is the completely free $\mathfrak{A}(2)$-algebra on a single generator of dimension n:

$$H^*(K(Z_2,n);Z_2) = U(\mathfrak{A}/B(n)).$$

The analogous result holds for $H^*(K(Z_p,n);Z_p)$, using computations of H. Cartan, Proc. Nat. Acad. Sci. 40 (1954), 704-707.

BIBLIOGRAPHY

[1] J. Milnor: "The Steenrod Algebra and Its Dual," <u>Annals of Math.</u>, 67 (1958), pp. 150-171.

CHAPTER III.

Embeddings of Spaces in Spheres.

In this chapter, we prove the non-embedding theorems of Thom and Hopf. Thom's theorem refers to an embedding of a compact space in a sphere, and Hopf's theorem to an embedding of an (n-1)-manifold in an n-sphere. In order to make duality work, we use Čech cohomology throughout this chapter.

§1. Thom's Theorem.

In this section, it is shown that if Y is a proper closed connected subspace of S^n, then

$$c(Sq^i): \ H^{n-2i}(Y;Z_2) \longrightarrow H^{n-i}(Y;Z_2), \qquad i > 0,$$

is zero. (See II §4 for the definition of c.)

1.1. LEMMA. All cup-products in $H^*(S^n,Y)$ are zero.

PROOF. Let $i^*: \ H^*(S^n,Y) \longrightarrow H^*(S^n)$. Let $u,v \in H^*(S,Y)$. Then $u \cup v \ = \ u \cup i^*v \ = \ i^*u \cup v \ = \ 0$ unless $u,v \in H^n(S^n,Y)$. In this case $u \cup v \in H^{2n}(S^n,Y)$. But, by duality, $H^{2n}(S^n,Y) \ \approx \ H_{-n}(S^n - Y) \ = \ 0$.

1.2. LEMMA. Let X be a compact Hausdorff space, and let $\{U_i\}$, $i \in I$, be a family of pairwise disjoint open subsets of X with union U. Then the maps

$$H^q(X,X - U_i) \longrightarrow H^q(X,X - U)$$

give a representation of $H^q(X,X - U)$ as a direct sum.

PROOF. Suppose first that I is finite. For any subspace Y of X, let $\bar{Y}$ denote its closure and $\dot{Y}$ its boundary. Let V be the disjoint topological union of the spaces $\bar{U}_i$, and let $W \subset V$ be the union of the spaces $\dot{U}_i$. Then (V,W) is a compact pair. The following diagram is commutative

$$\begin{array}{ccc} (\bar{U}_i, \dot{U}_i) & \longrightarrow & (V,W) \\ \downarrow & & \downarrow \\ (X, X - U_i) & \longleftarrow & (X, X - U) \ . \end{array}$$

Moreover, the vertical maps are relative homeomorphisms. We therefore get a commutative diagram

$$\begin{array}{ccc} H^q(\bar{U}_i, \dot{U}_i) & \longleftarrow & H^q(V,W) \\ \uparrow\approx & & \uparrow\approx \\ H^q(X, X - U_i) & \longrightarrow & H^q(X, X - U) \ . \end{array}$$

This in turn gives rise to a commutative diagram

$$\begin{array}{ccc} \Pi_i \ H^q(\bar{U}_i, \dot{U}_i) & \xleftarrow{\approx} & H^q(V,W) \\ \uparrow\approx & & \uparrow\approx \\ \Sigma_i \ H^q(X, X - U_i) & \longrightarrow & H^q(X, X - U) \ . \end{array}$$

So the lemma is proved when I is finite.

If I is infinite, we obtain the result by the continuity of Čech theory, taking limits over finite subsets of I.

1.3. LEMMA. If θ is any cohomology operation of one variable, such that

$$\theta\colon \ H^q(X) \longrightarrow H^n(X) \qquad (0 < q < n)$$

then $\theta\colon \ H^q(S^n, Y) \longrightarrow H^n(S^n, Y)$ is zero. (Note that the only axiom θ needs to satisfy is naturality with respect to mappings of spaces. θ need not be a homomorphism.)

PROOF. For any cohomology operation θ, with image in a positive dimension, $\theta(0) = 0$. The proof is as follows. Let X be any space, and let P be a point. Then we have the commutative diagram

$$\begin{array}{ccc} H^q(P) & \longrightarrow & H^q(X) \\ \downarrow\theta & & \downarrow\theta \\ H^n(P) & \longrightarrow & H^n(X) \end{array}$$

induced by the map $X \longrightarrow P$. Since $n > 0$, $H^n(P) = 0$, and so $\theta(0) = 0$, where $0 \in H^q(X)$.

So 1.2 shows that we have only to prove $\theta\colon H^q(S^n, S^n - U_i) \longrightarrow H^n(S^n, S^n - U_i)$ is zero, in order to prove our lemma. Now, we have the commutative diagram

$$\begin{array}{ccc} H^q(S^n, S^n - U_i) & \xrightarrow{\theta} & H^n(S^n, S^n - U_i) \\ \downarrow j^* & & \downarrow j^* \\ 0 = H^q(S^n) & \xrightarrow{\theta} & H^n(S^n) \end{array}$$

Since U_i is connected, we have by Alexander duality, $H^{n-1}(S^n - U_i) = 0$ and $H^n(S^n - U_i) = 0$. Therefore the vertical map on the right of the diagram is an isomorphism. This proves the lemma.

Let U be any neighbourhood of Y. Then there is a connected subcomplex K of S^n, which is a compact n-manifold with boundary L, such that $K \subset U$ and $Y \subset K - L$. We can construct K from the simplicial structure of S^n, by taking a fine subdivision. We can assume K is connected, since Y is connected. The set of such manifolds K, and the inclusion maps between them form an inverse system with limit Y. Therefore $H^*(Y)$ is the direct limit of the groups $H^*(K)$.

Let F be a field. We have the cup-product pairing

$$H^p(K,L;F) \otimes H^{n-p}(K;F) \longrightarrow H^n(K,L;F) \approx F.$$

Lefschetz duality tells us that the induced map

$$\alpha\colon\ H^p(K;F) \longrightarrow \mathrm{Hom}\,(H^{n-p}(K,L;F),F)$$

is an isomorphism. Let $x \in H^q(K;Z_2)$. We define a homomorphism

$$H^{n-q-i}(K,L;Z_2) \longrightarrow H^n(K,L;Z_2) \approx Z_2$$

by the formula $y \longrightarrow Sq^i y \cup x$. Let $Q^i x$ be the element of $H^{q+i}(K;Z_2)$ such that $\alpha(Q^i x)$ is the homomorphism. Then

$$Sq^i y \cup x = y \cup Q^i x.$$

Q^i is a homomorphism $Q^i\colon\ H^q(K;Z_2) \longrightarrow H^{q+i}(K;Z_2)$.

1.4. PROPOSITION. $Q^i = c(Sq^i)$ as a homomorphism $H^q(K;Z_2) \longrightarrow H^{q+i}(K;Z_2)$. (See II §4 for the definition of c.)

PROOF. We shall use Z_2 coefficients throughout this proof.

The proof is by induction on i. Obviously $Q^0 = 1$. Therefore $Q^0 = c(Sq^0)$. For any $x \in H^q(K)$ and $y \in H^{n-q-i}(K,L)$, we have, by definition,

$$
\begin{aligned}
y \cup (Q^i + \Sigma_{j=1}^{i-1} Q^j \, Sq^{i-j} + Sq^i)x
&= Sq^i y \cup x + \Sigma_{j=1}^{i-1} Sq^j y \cup Sq^{i-j} x + y \cup Sq^i x \\
&= Sq^i (y \cup x) \in H^n(K,L) \quad .
\end{aligned}
$$

We have the commutative diagram

$$
\begin{array}{ccc}
H^{n-i}(K,L) & \xrightarrow{\quad Sq^i \quad} & H^n(K,L) \\
\uparrow \approx & & \uparrow \approx \\
H^{n-i}(S^n, S^n - \text{Int } K) & \xrightarrow{Sq^i} & H^n(S^n, S^n - \text{Int } K) \quad .
\end{array}
$$

The vertical maps are excision isomorphisms. By 1.3, we have $Sq^i(y \cup x) = 0$ if $i > 0$. Therefore, from the computation above,

$$
\begin{aligned}
Q^i &= - \Sigma_j Q^j \, . \, Sq^{i-j} - Sq^i \\
&= - \Sigma_j c(Sq^j) \, . \, Sq^{i-j} - Sq^i && \text{by our induction hypothesis} \\
&= c(Sq^i) && \text{by the definition of } c \; .
\end{aligned}
$$

1.5. THEOREM. If the compact space Y can be embedded in S^n, then, for each $i > 0$, we have that

$$Q^i: \; H^{n-2i}(Y) \longrightarrow H^{n-i}(Y)$$

is zero. Equivalently, if a compact space Y is such that, for some r and $i > 0$,

$$Q^i: \; H^r(Y) \longrightarrow H^{r+i}(Y)$$

is not zero, then Y is not embeddable in S^{r+2i}.

PROOF. Suppose Y can be embedded in S^n. We construct a manifold K as described above. Let $y \in H^i(S^n, S^n - \text{Int } K)$. Then $Sq^i y = y^2 = 0$ by I §1 Axiom 3 and 1.1. We have the commutative diagram

$$
\begin{array}{ccc}
H^i(K,L) & \xrightarrow{\quad Sq^i \quad} & H^{2i}(K,L) \\
\uparrow \approx & & \uparrow \approx \\
H^i(S^n, S^n - \text{Int } K) & \xrightarrow{Sq^i} & H^{2i}(S^n, S^n - \text{Int } K) \quad .
\end{array}
$$

The vertical maps are excision isomorphisms. Since the lower horizontal map is zero, so is the upper one.

Let $x \in H^{n-2i}(K)$ and $y \in H^{i}(K,L)$. Then

$$y \cup Q^{i}x = Sq^{i}y \cup x = 0 .$$

By duality $Q^{i}x = 0$, since the above equation is true for all y.

1.6. LEMMA. Let x be a 1-dimensional cohomology class mod 2. Then $Q^{k}x = 0$ unless k has the form $2^{h} - 1$; if $k = 2^{h} - 1$, then $Q^{k}x = x^{2^{h}}$.

PROOF. This is proved by induction on k. It is obvious for $k = 0$. If $k > 0$, we have

$$0 = \sum_{i=0}^{k} Q^{i}Sq^{k-i}x = Q^{k}x + Q^{k-1}x^{2} .$$

Let $m\colon H^{*}(X) \otimes H^{*}(X) \longrightarrow H^{*}(X)$ be the cup-product, and let $\psi\colon \mathfrak{A}(2) \longrightarrow \mathfrak{A}(2) \otimes \mathfrak{A}(2)$ be the diagonal. Then

$$\begin{aligned} Q^{k-1}x^{2} &= c(Sq^{k-1})x^{2} \\ &= m[\psi(cSq^{k-1}) \,.\, x \otimes x] && \text{by II 1.3} \\ &= m[(c \times c)T\psi\, Sq^{k-1} \,.\, x \otimes x] && \text{by II §4} \\ &= m[\Sigma\, cSq^{i} \times cSq^{k-i-1} \,.\, x \otimes x] \\ &= \textstyle\sum_{i=0}^{k-1} Q^{i}x \,.\, Q^{k-i-1}x . \end{aligned}$$

The summation cancels out in pairs (mod 2), except for the middle term, if any. The middle term occurs when $i = k - i - 1$, and, by induction, is equal to $x^{2^{m}} . x^{2^{m}}$ if $i = 2^{m} - 1$ and is zero otherwise. So $Q^{k-1}x^{2} = x^{2^{m+1}}$ if $k = 2^{m+1} - 1$ and is zero otherwise. This proves the lemma.

1.7. THEOREM. If $1 < 2^{h} \leq n < 2^{h+1}$, then real projective n-space P_n cannot be embedded in a sphere of dimension less than 2^{h+1}.

PROOF. Let x be the generator of $H^{1}(P_n;Z_2)$. Then $Q^{2^{h}-1}x = x^{2^{h}} \neq 0$. By 1.5, the theorem follows.

1.7 was first proved for regular differentiable embeddings by using Stiefel-Whitney classes.

§2. Hopf's Theorem.

Let M be a closed $(n-1)$-manifold embedded in S^n. Applying Alexander duality with coefficients Z_2 and then with coefficients Z, we find that M is orientable and that M separates S^n into two open sets with closures A and B such that $A \cup B = S^n$. By duality no proper closed subset of M can separate S^n, and so $A \cap B = M$. Applying duality to A and then to B, we see that

$$H^r(A) = H^r(B) = 0 \qquad (r \geq n-1)$$

for any ring of coefficients. We have the following theorem due to Hopf.

2.1. THEOREM. Under the above hypotheses, the inclusion maps $i\colon M \subset A$ and $j\colon M \subset B$ induce a representation of $H^q(M)$ as a direct sum

$$H^q(M) = i^*H^q(A) + j^*H^q(B) \quad \text{for} \quad 0 < q < n-1.$$

Here i^* and j^* are monomorphisms. Using a field of coefficients F, and the identification $H^{n-1}(M) = F$, cup-products in M give an isomorphism

$$i^*H^q(A) \approx \text{Hom}\,(j^*H^{n-q-1}(B), F) \quad \text{for } 0 < q < n-1.$$

PROOF. The first statement follows immediately from the Mayer-Vietoris sequence. Since $H^{n-1}(A) = H^{n-1}(B) = 0$, cup-products in A or B with values in dimension $(n-1)$ are zero. The rest of the theorem follows by Poincaré duality.

2.2. COROLLARY. If $n \geq 2$, then real projective n-space cannot be embedded in S^{n+1}.

2.3. LEMMA. Let $x \in H^r(M;Z_2)$ and let $r + k = n - 1$, then $Sq^k x = 0$.

PROOF. Let $x = i^*a + j^*b$, where $a \in H^r(A)$ and $b \in H^r(B)$. The lemma follows by naturality since $H^{n-1}(A) = H^{n-1}(B) = 0$.

Let $Q^k = c(Sq^k)$ as in §1. Let $x \in H^r(A;Z_2)$. Suppose the action of the Steenrod algebra $\mathfrak{A}(2)$ on $H^*(B;Z_2)$ is known. Then $Sq^k x$ is determined by the following theorem.

2.4. THEOREM. Let $s = n - 1 - r - k$ and let $y \in H^s(B;Z_2)$, $x \in H^r(A;Z_2)$; then

$$i^*Sq^k x \cup j^*y = i^*x \cup j^*Q^k y .$$

PROOF. The theorem is proved by induction on k. It is obvious for $k = 0$.

By 2.3, $Sq^k(i^*x \cup j^*y) = 0$ if $k > 0$. So by the Cartan formula

$$0 = \sum_{m=0}^{k} Sq^m i^*x \cup Sq^{k-m} j^*y$$

$$= \sum_{m=0}^{k-1} i^*x \cup Q^m Sq^{k-m} j^*y + i^*Sq^k x \cup j^*y$$

by our induction hypothesis

$$= i^*x \cup j^*Q^k y + i^*Sq^k x \cup j^*y$$

by the definition of Q^k in II §4.

Therefore $i^*x \cup j^*Q^k y = i^*Sq^k x \cup j^*y$.

BIBLIOGRAPHY

Thom, R., "Espaces fibrés en spheres et carrés de Steenrod," Ann. Sci. Ecole Normale Sup. 69 (1952), 109-182.

CHAPTER IV.

The Cohomology of Classical Groups and Stiefel Manifolds.

In this chapter, we find the cohomology rings of the real, complex and quaternionic Stiefel manifolds. We also obtain the Pontrjagin rings of the orthogonal, unitary and symplectic groups and of the special orthogonal and special unitary groups. The method is to obtain a cellular decomposition of the Stiefel manifolds (following [1] and [2]). We then find the action of the Steenrod algebra $\mathfrak{A}(2)$ in the cohomology rings of the real Stiefel manifolds. Using this information, we obtain an upper bound on the possible number of linearly independent vector fields on a sphere.

§1. Definitions.

Let $F = F_d$ be the real or complex numbers or the quaternions, according as $d = 1, 2$ or 4. Let $V = F^n$ be the n-dimensional vector space over F, consisting of column vectors with entries in F. We write scalars on the right. Let u_i be the column vector with 1 in the ith row and zero elsewhere. Let $x = \Sigma_i u_i x_i \in V$, where $x_i \in F$ and let $y = \Sigma_i u_i y_i$. We define the scalar product $< x,y > = \Sigma_i \bar{x}_i y_i$, where $\bar{x}_i$ is the conjugate of x_i. Then $< x,y\lambda > = < x,y >\lambda$ if $\lambda \in F$; $< x,y_1 + y_2> = < x,y_1 > + < x,y_2 >$; $< x_1 + x_2,y > = < x_1,y> + < x_2,y >$; and $< x,y > = \overline{< y,x >}$. We embed F^n in F^{n+1} by putting the last coordinate equal to zero.

Let $G(n)$ be the group of transformations of V which preserve scalar products. That is, $A \in G(n)$ if and only if $< Ax,Ay > = < x,y >$ for all $x,y \in V$. If A is represented by the $n \times n$-matrix $[a_{ij}]$ multiplying column vectors on the left, then $A \in G(n)$ if and only if $\bar{A}^t A = I$. $G(n)$ is the orthogonal, unitary or symplectic group according as $d = 1, 2$ or 4. We have an embedding $G(n) \subset G(n+1)$ induced by the embedding $F^n \subset F^{n+1}$. The matrix $A \in G(n)$ corresponds to the matrix

$$\begin{pmatrix} A & 0 \\ 0 & 1 \end{pmatrix} \in G(n+1)$$

We write $G(0) = I$.

The Stiefel manifold $G(n,k)$ is the manifold of left cosets $G(n)/G(k)$. Let $G'(n,k)$ be the manifold of $(n-k)$-frames in n-space. The mapping $G(n) \longrightarrow G'(n,k)$, which selects the last $(n-k)$ columns of a matrix as the $(n-k)$ vectors of an $(n-k)$-frame, induces a map $G(n,k) \longrightarrow G'(n,k)$ which is obviously onto. If two matrices A and B in $G(n)$ have the same last $(n-k)$ columns, then $A^{-1}B \in G(k)$. Therefore the map $G(n,k) \longrightarrow G'(n,k)$ is a homeomorphism, and we can identify the two spaces. Now $G'(n,n-1)$ is the manifold of unit vectors in V. Therefore

1.1. $G(n,n-1)$ is homeomorphic to $S^{nd-1} \subset V = F^n$ by the map which selects the last column of a matrix.

1.2. Definition of φ. Let S^{nd-1} be the sphere of unit vectors in $V = F^n$. Then S^{d-1} is the sphere of scalars of unit norm in F. We construct a mapping

$$\varphi: S^{nd-1} \times S^{d-1} \longrightarrow G(n)$$

by letting $\varphi(x,\lambda)$ be the transformation which keeps y fixed if $\langle x,y \rangle = 0$, and which sends x to $x\lambda$. That is

$$\varphi(x,\lambda)y = x(\lambda-1)\langle x,y \rangle + y \quad \text{or}$$

$$\varphi(x,\lambda)_{ij} = x_i(\lambda-1)\bar{x}_j + \delta_{ij} \qquad \text{in matrix notation.}$$

If $m < n$ we have an inclusion $S^{md-1} \longrightarrow S^{nd-1}$, induced by the inclusion $F^m \longrightarrow F^n$. This induces a further inclusion

$$S^{md-1} \times S^{d-1} \longrightarrow S^{nd-1} \times S^{d-1}.$$

The following diagram is obviously commutative

$$\begin{array}{ccc} S^{md-1} \times S^{d-1} & \longrightarrow & S^{nd-1} \times S^{d-1} \\ \downarrow \varphi & & \downarrow \varphi \\ G(m) & \longrightarrow & G(n) \end{array}.$$

1.3. <u>Definition of Q_n</u>. Let Q_n be the quotient space of $S^{nd-1} \times S^{d-1}$ induced by φ. It is the set of pairs $(x,\lambda) \in S^{nd-1} \times S^{d-1}$ under the identifications $(x,\lambda) = (x\nu, \nu^{-1}\lambda\nu)$ where $\nu \in S^{d-1}$ and $(x,1) = (y,1)$. That these are the only identifications is easily seen by looking at the fixed point set of $\varphi(x,\lambda)$. Let Q_0 be a single point. We embed Q_0 in Q_n $(n \geq 1)$ by sending Q_0 to the equivalence class of $(x,1)$.

If $n > m \geq 1$, we have an embedding $Q_m \longrightarrow Q_n$ induced by $S^{md-1} \times S^{d-1} \longrightarrow S^{nd-1} \times S^{d-1}$. By the commutativity of the previous diagram, we have another commutative diagram

$$\begin{array}{ccc} Q_m & \longrightarrow & Q_n \\ \downarrow & & \downarrow \\ G(m) & \longrightarrow & G(n) \end{array}$$

whenever $n > m \geq 0$.

Q_m is a compact Hausdorff space and so the vertical maps are embeddings. We identify Q_m with its embedded image in $G(n)$ $(n \geq m)$. Under the identification Q_0 becomes the identity of $G(n)$.

Let $E^{(n-1)d}$ be the ball consisting of all vectors $x \in S^{nd-1} \subset V = F^n$, with x_n real and $x_n \geq 0$. Then x_n is determined by $x_1,\ldots,x_{n-1}$. Let $f_n\colon E^{(n-1)d} \longrightarrow S^{nd-1}$ be the inclusion map. Let $g\colon (E^{d-1}, S^{d-2}) \longrightarrow (S^{d-1}, 1)$ be the usual relative homeomorphism $(S^{-1} = \emptyset)$. Let

$$h_n\colon\ E^{nd-1} \longrightarrow Q_n \qquad (n \geq 1)$$

be the composition

$$E^{nd-1} = E^{(n-1)d} \times E^{d-1} \xrightarrow{f_n \times g} S^{nd-1} \times S^{d-1} \longrightarrow Q_n .$$

Let S^{nd-2} be the boundary of E^{nd-1}.

1.4. LEMMA. The map h_n defines a relative homeomorphism $h_n\colon (E^{nd-1}, S^{nd-2}) \longrightarrow (Q_n, Q_{n-1})$ if $n \geq 1$. Therefore Q_n is a CW complex with a 0-cell Q_0 and with an $(md-1)$-cell for each m such that $1 \leq m \leq n$.

PROOF. $(f_n \times g)S^{nd-2}$ consists of points of the form $(x,\lambda) \in S^{nd-1} \times S^{d-1}$ where $x_n = 0$ or $\lambda = 1$. Therefore $h(S^{nd-2}) \subset Q_{n-1}$.

In any equivalence class $\{(x,\lambda)\} \in S^{nd-1}$ we can choose a representative (x,λ) so that x_n is real and $x_n \geq 0$. Moreover, if $\lambda \neq 1$ and $x_n > 0$, this representative is unique. This proves that h_n: $(E^{nd-1}, S^{nd-2}) \longrightarrow (Q_n, Q_{n-1})$ is a relative homeomorphism.

The rest of the lemma follows by induction on n.

1.5. DEFINITION. Let μ be the multiplication in $G(n)$. Let π: $G(n) \longrightarrow G(n,k)$ be the standard projection. A <u>normal cell</u> of $G(n,k)$ is a map (or the image of a map) of the form

$$E^{i_1 d-1} \times \dots \times E^{i_r d-1} \xrightarrow{h \times \dots \times h} Q_{i_1} \times \dots \times Q_{i_r} \xrightarrow{\pi\mu} G(n,k)$$

where $n \geq i_1 > i_2 > \dots > i_r > k$. We denote such a cell by $(i_1, \dots, i_r | n,k)$ or simply by $(i_1, \dots, i_r)$ if this will cause no confusion. The cells of Q_n (other than Q_0), described in 1.4, may be identified with the normal cells $(m|n,0)$ where $n \geq m > 0$. We denote such a cell of Q_n by (m).

By μ we shall also denote the action of $G(n)$ by left translation on $G(n,k)$ $(n \geq k \geq 0)$.

§2. <u>The Cellular Structure of the Stiefel Manifolds</u>.

In this section we shall prove the following pivotal theorem.

2.1. THEOREM. $G(n,k)$ is a CW complex, whose cells are the normal cells (see 1.5) and the 0-cell $\pi(I)$. The map

$$\mu: \; Q_n \times G(n-1,k) \longrightarrow G(n,k) \qquad (k < n)$$

is cellular and induces an epimorphism of chain complexes.

Before proving the theorem, we state and prove a corollary.

2.2. COROLLARY. If $m \leq n$ and $\ell \leq k$, then the induced map $G(m,\ell) \longrightarrow G(n,k)$ is cellular. This map sends the normal cell $(i_1, \dots, i_r | m,\ell)$ to the normal cell $(i_1, \dots, i_r | n,k)$ if $i_r > k$; to $(i_1, \dots, i_{r-1} | n,k)$ if $d = 1$ and $i_{r-1} > k \geq i_r = 1 > \ell = 0$; and degenerately otherwise.

PROOF. This follows immediately from 2.1 and the definition 1.5 of normal cells.

We now begin the proof of 2.1.

Let us denote by α: $(Q_n, Q_{n-1}) \longrightarrow (S^{nd-1}, u_n)$ the composition

$$(Q_n, Q_{n-1}) \xrightarrow{\pi} (G(n,n-1), G(n-1,n-1)) \longrightarrow (S^{nd-1}, u_n)$$

where the map on the right is the homeomorphism of 1.1.

2.3. LEMMA. The map α: $(Q_n, Q_{n-1}) \longrightarrow (S^{nd-1}, u_n)$ is a relative homeomorphism.

PROOF. α: $Q_n \longrightarrow S^{nd-1}$ sends $\{(x,\lambda)\}$ $(x \in S^{nd-1}, \lambda \in S^{d-1})$ to $x(\lambda - 1)\bar{x}_n + u_n$ by 1.2. The inverse image of u_n under α is Q_{n-1}, for if $x(\lambda - 1)\bar{x}_n + u_n = u_n$, then $\lambda = 1$ or $x_n = 0$.

Suppose we are given $y \in S^{nd-1} \subset V = F^n$ such that $y \neq u_n$. We must show that there is exactly one element $(x,\lambda) \in S^{nd-1} \times S^{d-1}$ with x_n real and $x_n > 0$, and $\lambda \neq 1$, such that $x(\lambda - 1)x_n = y - u_n$.

In the real d-dimensional space F, $(y_n - 1)$ lies in the closed ball bounded by the sphere of scalars of the form $(\lambda - 1)$, where $|\lambda| = 1$. Moreover, $(y_n - 1) \neq 0$. So, projecting from the origin in the real d-dimensional space F, we can solve uniquely the equation $x_n^2(\lambda - 1) = (y_n - 1)$, for x_n real, $x_n > 0$, $|\lambda| = 1$ and $\lambda \neq 1$. Knowing λ and x_n, x_i is determined uniquely for $1 \leq i \leq n-1$. We now have $x(\lambda - 1)x_n = y - u_n$, x_n real, $|\lambda| = 1$ and $\lambda \neq 1$.

We have to check that $\langle x,x \rangle = 1$. On evaluating the scalar product of each side of the above equation with itself, we find

$$\langle x,x \rangle (2 - \lambda - \bar{\lambda})x_n^2 = 2 - y_n - \bar{y}_n .$$

Also, we know that $x_n^2(\lambda - 1) = (y_n - 1)$. Hence

$$\langle x,x \rangle (2 - \lambda - \bar{\lambda})x_n^2 = (2 - \lambda - \bar{\lambda})x_n^2 .$$

Since $|\lambda| = 1$ and $\lambda \neq 1$, we have $(2 - \lambda - \bar{\lambda}) \neq 0$. Since also, $x_n \neq 0$, we deduce that $\langle x,x \rangle = 1$.

2.4. PROPOSITION. If $n > k \geq 0$, then

$$\mu: (Q_n \times G(n-1,k), Q_{n-1} \times G(n-1,k)) \longrightarrow (G(n,k), G(n-1,k))$$

is a relative homeomorphism and maps $Q_n \times G(n-1,k)$ onto $G(n,k)$.

PROOF. The inverse image of $G(n-1,k)$ is $Q_{n-1} \times G(n-1,k)$. To see this, let $A \in Q_n, B \in G(n-1)$ and suppose $ABG(k) \subset G(n-1)$. Then $A \in G(n-1)$. On projecting into $G(n,n-1)$, we see by 2.3 that $Q_n \cap G(n-1) = Q_{n-1}$. Therefore $A \in Q_{n-1}$.

μ is one-to-one on $(Q_n - Q_{n-1}) \times G(n-1,k)$. To see this, let

$A, C \in Q_n - Q_{n-1}$, and let $B, D \in G(n-1)$ and suppose that $ABG(k) = CDG(k)$. Then $AG(n-1) = CG(n-1)$. On projecting into $G(n, n-1)$, we see by 2.3 that $A = C$. Therefore $BG(k) = DG(k)$.

μ maps $Q_n \times G(n-1,k)$ onto $G(n,k)$. To see this, let $A \in G(n)$. By 2.3, there is an element $C \in Q_n$ such that $CG(n-1) = AG(n-1)$. Therefore there is an element $D \in G(n-1)$ such that $A = CD$.

2.5. LEMMA. Let $x \in S^{nd-1} \subset V = F^n$ be a unit vector, let $\lambda \in S^{d-1} \subset F$ be a unit scalar and let $A \in G(n)$. Then $A\,\varphi(x,\lambda)A^{-1} = \varphi(Ax,\lambda)$. (See 1.2 for definition of φ.)

PROOF. By definition, $\varphi(Ax,\lambda)$ is the transformation which keeps y fixed if $\langle Ax, y \rangle = 0$ and which sends Ax to $Ax\lambda$. The lemma follows.

2.6. PROPOSITION. $\mu(Q_m \times Q_m) = \mu(Q_m \times Q_{m-1}) \subset G(m)$ for $m \geq 1$.

PROOF. We shall reduce the case where m is arbitrary to the case where $m = 2$. We therefore begin by checking the proposition for $m = 1$ and $m = 2$.

If $m = 1$, then $Q_1 = G(1)$. We see this from 2.4 by putting $n = 1$ and $k = 0$. The proposition follows since $G(1) = \mu(Q_1 \times Q_0) \subset \mu(Q_1 \times Q_1) \subset G(1)$ (recall from 1.3 that Q_0 is the identity element).

If $m = 2$, then $\mu(Q_2 \times Q_1) = G(2)$. We see this from 2.4, by putting $n = 2$ and $k = 0$, and recalling that $Q_1 = G(1)$ from the previous paragraph. So

$$G(2) = \mu(Q_2 \times Q_1) \subset \mu(Q_2 \times Q_2) \subset G(2) .$$

The proposition follows.

Now let $x, y \in S^{md-1} \subset V = F^m$ and let $\lambda, \nu \in S^{d-1} \subset F$ be arbitrary elements. Then $\varphi(x,\lambda)$ and $\varphi(y,\nu)$ are arbitrary elements of Q_m. We must show that $\varphi(x,\lambda)\varphi(y,\nu) \in Q_m Q_{m-1}$.

Let W be a 2-dimensional subspace of V containing x and y. Using the inclusion $F^r \subset F^{r+1}$ of §1, we have the sequence

$$0 \subseteq W \cap F^1 \ldots \subseteq W \cap F^m = W$$

of vector spaces over F, increasing by at most one dimension at a time. We choose an integer r, such that $1 \leq r < m$ and $W \cap F^r$ is 1-dimensional. Let $A \in F(m)$ map W onto F^2 so that $A(W \cap F^r) = F^1$. Let $Ax = x' \in F^2$ and $Ay = y' \in F^2$. Then, by 2.5,

$$A\,\varphi(x,\lambda)\varphi(y,\nu)A^{-1} = \varphi(x',\lambda)\varphi(y',\nu) .$$

Since the proposition is true for $m = 2$,

$$\varphi(x',\lambda)\varphi(y',\nu) \in Q_2 Q_1 .$$

Therefore

$$A\,\varphi(x,\lambda)\varphi(y,\nu)A^{-1} = \varphi(x_1,\lambda_1)\varphi(y_1,\nu_1)$$

where $x_1 \in S^{2d-1} \subset F^2$, $y_1 \in S^{d-1} \subset F^1$, $\lambda_1, \nu_1 \in S^{d-1} \subset F$. Again using 2.5, we see that

$$\varphi(x,\lambda)\varphi(y,\nu) = \varphi(A^{-1}x_1,\lambda_1)\varphi(A^{-1}y_1,\nu_1) .$$

By our choice of A, $A^{-1}y_1 \in W \cap F^r \leq F^{m-1}$. Therefore

$$\varphi(x,\lambda)\varphi(y,\nu) \in Q_m Q_{m-1} .$$

This completes the proof of the proposition.

PROOF of 2.1. We denote 2.1 when $n = m$ by 2.1(m). We shall prove 2.1 and the following two statements together by induction on n.

2.7(n). Let $n \geq i_1, \ldots, i_r > 0$. Then $\pi(Q_{i_1} \ldots Q_{i_r})$ is contained in the $(\Sigma_{s=1}^{r}(i_s d - 1))$-skeleton of $G(n,k)$.

2.8(n). $\mu\colon\ Q_n \times G(n,k) \longrightarrow G(n,k)$ is cellular.

When $n = k$, all the assertions are obvious, for then $G(n,k)$ is the point $\pi(I)$. Suppose $n > k$ and that 2.1(n - 1), 2.7(n - 1) and 2.8(n - 1) are true.

By 2.8(n - 1), $\mu\colon\ Q_{n-1} \times G(n-1,k) \longrightarrow G(n-1,k)$ is cellular. By 2.4 and 2.1(n - 1), $G(n,k)$ therefore has a CW structure such that the map $\mu\colon\ Q_n \times G(n-1,k) \longrightarrow G(n,k)$ is cellular. By 2.4, 1.4 and 2.1(n-1), the cells of $G(n,k)$ other than those in $G(n-1,k)$, are of the form $\mu((n) \times (i_1,\ldots,i_r))$ where $n - 1 \geq i_1 > \ldots > i_r > k$. Now $\mu((n) \times (i_1,\ldots,i_r)) = (n,i_1,\ldots,i_r)$ by 1.5, and this is a normal cell of $G(n,k)$. Therefore μ induces an epimorphism of chain complexes and 2.1(n) follows.

We now prove 2.7(n). By 2.5, if $A \in G(n)$, then $AQ_m = Q_mA$. Therefore $Q_jQ_m = Q_mQ_j$ $(0 \leq j < m)$. So, by 2.6 and 2.7(n - 1), we can assume without loss of generality that in the hypotheses of 2.7(n), $n = i_1 > i_2 \dots > i_r > k$. Now $\pi(Q_{i_2} \dots Q_{i_r}) \subset G(n-1,k)$. Therefore by 2.7(n - 1), $\pi(Q_{i_2} \dots Q_{i_r})$ is contained in the $(\Sigma_{s=2}^{r}(i_s d - 1))$-skeleton of $G(n-1,k)$. By 2.1(n),

$$\mu: \ Q_n \times G(n-1,k) \longrightarrow G(n,k)$$

is cellular. Since Q_n has dimension (nd-1), 2.7(n) follows.

We now prove 2.8(n). Since Q_0 is the identity, μ is cellular on $Q_0 \times G(n,k)$. By 2.1(n), μ is cellular on $Q_n \times G(n-1,k)$. We have only to check that μ is cellular on cells of the form $(t) \times (n,i_1,\dots,i_r)$ where $n \geq t > 0$ and $n > i_1 \dots > i_t > k$ (see 2.1(n)). Now $\mu((t) \times (n,i_1,\dots,i_r)) \subset \pi(Q_tQ_nQ_{i_1} \dots Q_{i_r})$ and our assertion follows from 2.7(n). This completes our proof of 2.1, 2.7 and 2.8.

§3. The Pontrjagin Rings of the Groups G(n).

3.1. Throughout the remainder of this chapter, all chain and cochain complexes and all homology and cohomology groups will be taken with coefficients R, where R is a commutative ring with a unit if $d = 2$ or 4, and $R = Z_2$ if $d = 1$.

The aim of this section is to find the Pontrjagin rings of the orthogonal group O(n), the unitary groups U(n) and the symplectic group Sp(n) (i.e., G(n) in the cases $d = 1,2$ and 4 respectively). That is, we want a description of the map

$$H_*(G(n);R) \otimes H_*(G(n);R) \longrightarrow H_*(G(n);R)$$

induced by the multiplication $G(n) \times G(n) \longrightarrow G(n)$.

3.2. LEMMA. If $d = 1$, Q_n is the disjoint union of the point Q_0 and the real projective space P^{n-1}. The embedding of Q_{n-1} in Q_n $(n \geq 2)$ corresponds to the usual embedding of P^{n-2} in P^{n-1}. $Q_1 = G(1)$ consists of two points, the 1×1 matrices I and -I. $(Q_n - Q_0)$ consists entirely of matrices of determinant -1.

If $d = 2$, Q_n is the suspension of the complex projective space CP^{n-1}, with the two suspension points identified. The embedding of Q_{n-1}

in Q_n $(n \geq 2)$ corresponds to the usual embedding of CP^{n-2} in CP^{n-1}.

PROOF. By 1.3, if $d = 1$ or 2, Q_n is the set of pairs $(x,\lambda) \in S^{nd-1} \times S^{d-1} \subset F^n \times F$ under the identifications $(x\nu, \lambda) = (x,\lambda)$ if $\nu \in S^{d-1} \subset F$, and $(x,1) = (y,1)$ if $y \in S^{nd-1} \subset F^n$. The second part of the lemma follows.

If $d = 1$, then for any pair (x,λ), if $\lambda \neq 1$ then $\lambda = -1$. The space Q_n therefore reduces to the disjoint union of Q_0 and the set of points $(x,-1)$ under the identifications $(x,-1) = (-x,-1)$.

So, if $d = 1$, Q_1 consists of two points and so does $G(1)$. Since $Q_1 \subset G(1)$, $Q_1 = G(1)$ and $(Q_1 - Q_0)$ is the matrix $(-I) \in G(1)$. This matrix has determinant -1. By connectedness, all matrices in $(Q_n - Q_0)$ therefore have determinant -1. (All matrices in $O(n)$ have determinant ± 1.) This completes the proof of the lemma.

The boundary of each cell in Q_n is algebraically zero. If $d = 2$ or 4, this follows from 1.4 for dimensional reasons. If $d = 1$, it follows from 3.1 and 3.2.

By 2.1 $\mu\colon Q_n \times G(n-1,k) \longrightarrow G(n,k)$ is an epimorphism of chain complexes. By induction on n, the boundaries of the cells of $G(n,k)$ are algebraically zero. Therefore there are no boundaries in $G(n,k)$ and all chains are cycles.

3.3. DEFINITION. If $(i_1,\ldots,i_r|n,k)$ is a normal cell (see 1.5), we denote its homology class by $[i_1,\ldots,i_r|n,k]$ or $[i_1,\ldots,i_r]$. We denote by $\{i_1,\ldots,i_r|n,k\}$ or $\{i_1,\ldots,i_r\}$, the cohomology class of $G(n,k)$ which assigns the value 1 to the normal cell $(i_1,\ldots,i_r)$ and zero to all other cells. We call these homology and cohomology classes <u>normal classes</u>. We denote the homology class of $\pi(I)$ by $\underline{1}$. We denote by $\bar{1}$, the cohomology class which assigns the value 1 to $\pi(I)$ and the value 0 to all other cells. We call $\underline{1}$ and $\bar{1}$ <u>unit classes</u>.

The following lemma is an immediate consequence of 2.1.

3.4. LEMMA. $H_*(G(n,k);R)$ is the free R-module on the unit class $\underline{1}$ and on the normal classes $[i_1,\ldots,i_r|n,k]$. If $n \leq m$ and $k \leq \ell$, we have a map $G(n,k) \longrightarrow G(n,\ell)$ which sends $[i_1,\ldots,i_r|n,k]$ to

$[i_1,\ldots,i_r|m,\ell]$ if $i > \ell$; to $[i_1,\ldots,i_{r-1}|m,\ell]$ if $d = 1$ and $i_{r-1} > \ell \geq i_r = 1 > k = 0$; and to zero otherwise. The map

$$\mu_*\colon\ H_*(Q_n \times G(n-1,k);R) \longrightarrow H_*(G(n,k);R)$$

is an epimorphism.

3.5. THEOREM. The Pontrjagin ring of $G(n)$ is the commutative, associative algebra over R with unit element $\underline{1}$ and generated by the normal classes $[i]$ of dimension $(id-1)$, where $n \geq i > 0$, subject to the relations

$$\begin{aligned} [i]\,[j] &= -[j]\,[i] && \text{if } i \neq j \\ [1]\,[1] &= \underline{1} && \text{if } d = 1 \\ [i]\,[i] &= 0 && \text{if } i > 1 \text{ or } d > 1\,. \end{aligned}$$

The normal class $[i_1,i_2,\ldots,i_r|n,0] = [i_1]\,[i_2]\ldots[i_r]$.

PROOF. Let $1 \leq j < i$. We have the diagram

$$\begin{array}{ccccc} E^{id-1} \times E^{jd-1} & \xrightarrow{\psi} & E^{jd-1} \times E^{id-1} & \xrightarrow{\theta} & E^{jd-1} \times E^{id-1} \\ \Big\downarrow h_i \times h_j & & & & \Big\downarrow h_j \times h_i \\ Q_i \times Q_j & \xrightarrow{\mu} & G(n) & \xleftarrow{\mu} & Q_j \times Q_i\,. \end{array}$$

Here $\psi(x,y) = (y,x)$, h_i and h_j are the maps of 1.4, μ is the multiplication and θ is defined as follows. Let $E^{id-1} = E^{(i-1)d} \times E^{d-1}$, where $E^{(i-1)d} \subset S^{id-1} \subset V = F^i$ is the set of all unit vectors x with x_n real and $x_n \geq 0$. $E^{(i-1)d}$ is invariant under $G(j)$ since $j < i$. We define $\theta(x,y_1,y_2)$, where $x \in E^{jd-1}$, $y_1 \in E^{(i-1)d}$ and $y_2 \in E^{d-1}$, to be $(x,(h_jx)^{-1}y_1,y_2)$. This definition is meaningful since $h_jx \in Q_j \subset G(j)$. By 2.5 and the definition of h_i, the diagram is commutative.

We now find the degree of the maps ψ and θ. If $d = 2$ or 4, both factors have odd dimension, and so ψ has degree -1. If $d = 1$, we are working mod 2 and signs don't matter. Also, θ has degree $(-1)^d$. To see this, let $f\colon E^{jd-1} \times I \longrightarrow E^{jd-1}$ be a contraction of E^{jd-1} onto a point z. This gives a homotopy of θ

$$(x,y_1,y_2,t) \longrightarrow (x,(h_jf(x,t))^{-1}y_1,y_2)$$

which shows that θ is homotopic to the map

$$(x,y_1,y_2) \longrightarrow (x,h_j(z)^{-1}y_1,y_2)$$

by a homotopy which is a homeomorphism at each time t. Therefore θ has the same degree as $h_j(z)$. Each element of $U(n)$ or $Sp(n)$ has degree 1, since these groups are connected. If $d = 1$, $h_j(z)$ has degree -1 by 1.4 and 3.2. So θ has degree $(-1)^d$.

Since the normal cell $(i) \subset Q_i$, and, by 2.6, $Q_iQ_i = Q_iQ_{i-1}$, we have $(i)(i) \subset Q_iQ_{i-1}$. If $i > 1$, then by 2.7, Q_iQ_{i-1} is contained in the skeleton of $G(n)$ of dimension $(2i - 1)d - 2$ which is less than $2(id - 1)$. Therefore $[i][i]$ is zero for dimensional reasons, if $i > 1$ If $i = 1$, then $(i)(i) \subset Q_1$, which has dimension $d - 1$. If $d > 1$, then $[1][1]$ is zero for dimensional reasons.

If $d = 1$, then by 3.2 there are two 0-cells, namely the 1×1 matrices I and $-I$. (1) is the 0-cell $-I$. Therefore $[1][1] = \underline{1}$.

In order to complete the proof of 3.5, it only remains to be shown that the classes $\underline{1}$ and $[i_1][i_2] \ldots [i_r]$, where $n \geq i_1 > i_2 \ldots > i_r > 0$ form a free basis for $H_*(G(n);R)$. This follows from 3.4, since the definition 1.5 of normal cells shows that

$$[i_1][i_2] \ldots [i_r] = [i_1,\ldots,i_r|n,0] .$$

The map $\mu\colon G(n) \times G(n,k) \longrightarrow G(n,k)$ gives $H_*(G(n,k);R)$ the structure of a module over Pontrjagin ring $H_*(G(n);R)$.

3.6. THEOREM. $H_*(G(n,k);R)$ is a module over $H_*(G(n);R)$ on a single generator $\underline{1}$ $(n > k \geq 1)$. The defining relations for this module are

$$[i]\,\underline{1} = 0 \quad \text{if } k \geq i > 0 \quad \text{if } d \neq 1$$

$$[i]\,\underline{1} = 0 \quad \text{if } k \geq i > 1 \quad \text{if } d = 1$$

$$[1]\,\underline{1} = \underline{1} \quad \text{if } d = 1 .$$

The normal class $[i_1,i_2,\ldots,i_r|n,k] = [i_1][i_2] \ldots [i_r]\,\underline{1}$.

PROOF. This follows immediately from 3.4 and 3.5 .

§4. The Cohomology Rings $H^*(G(n,k);R)$.

We begin this section by reminding the reader of our assumption 3.1 on R.

We shall compute the cohomology ring of $G(n,k)$ by induction on n and by use of the monomorphism (see 3.4)

$$\mu^*: \; H^*(G(n,k);R) \longrightarrow H^*(Q_n \times G(n-1,k);R) \quad \text{where} \quad n > k.$$

If $d = 1$, we write $O(n,k) = G(n,k)$; if $d = 2$, $U(n,k) = G(n,k)$; if $d = 4$, $Sp(n,k) = G(n,k)$.

4.1. NOTATION. We extend the notation of 3.3 as follows. Let $i_1,\ldots,i_r$ be a set of integers all greater than k, where $k \geq 0$ if $d = 2$ or 4, and $k \geq 1$ if $d = 1$. Let $\{i_1,\ldots,i_r\}$ be the zero cohomology class of $G(n,k)$ if $i_s > n$ for some s such that $1 \leq s \leq r$, or if $i_s = i_t$ where $1 \leq s < t \leq r$. Otherwise let $\{i_1,\ldots,i_r\}$ denote the product of the normal class of $G(n,k)$ obtained on permuting $i_1,\ldots,i_r$, and the sign of the permutation. (Recall that (i_r) is a cell of dimension $(i_r d - 1)$, which is odd if $d = 2$ or 4, while if $d = 1$, our ring is Z_2. Therefore this notation is consistent with the usual convention for sign-changing.) Curly brackets with a space between them $\{\ \}$, should be interpreted as $\bar{1}$. We also use the symbols $\{b\}$ and $\bar{1}$, where $0 < b \leq n$ to denote the images of these classes under the map $H^*(G(n);R) \longrightarrow H^*(Q_n;R)$.

4.2. LEMMA. a) Let $n \geq k \geq 1$. Under the monomorphism (see 3.4)

$$\mu^*: \; H^*(O(n,k);Z_2) \longrightarrow H^*(Q_n \times O(n-1,k);Z_2),$$

we have, in the notation of 4.1,

$$\mu^*\{b_1,\ldots,b_r\} = \bar{1} \times \{b_1,\ldots,b_r\} + \{1\} \times \{b_1,\ldots,b_r\} + \Sigma_{i=1}^r \{b_i\} \times \{b_1,\ldots,b_{i-1},b_{i+1},\ldots,b_r\}.$$

b) Let $n \geq k > 0$. Let $d = 2$ or 4. Under the monomorphism

$$\mu^*: \; H^*(G(n,k);R) \longrightarrow H^*(Q_n \times G(n-1,k);R),$$

we have, in the notation of 4.1,

$$\mu^*\{b_1,\ldots,b_r\} = \bar{1} \times \{b_1,\ldots,b_r\} + \Sigma_{i=1}^r (-1)^{r+i}\{b_i\} \times \times \{b_1,\ldots,b_{i-1},b_{i+1},\ldots,b_r\}.$$

PROOF. Without loss of generality, we may assume that $n \geq b_1 > \ldots > b_r > k$, since interchanging two adjacent b's multiplies both sides of the equation by -1, and if two of the b's are equal, both sides of the equation are zero.

We prove the theorem by evaluating both sides of the equation on cells of $Q_n \times G(n-1,k)$. The value of the left-hand side is calculated with the help of 3.5 and 3.6. When evaluating the right hand side, we must use the sign change $< c_1 \times c_2, h_1 \times h_2 > \; = \; (-1)^{pq} < c_1, j_1 > < c_2, h_2 >$, where $c_1 \in H^p(X)$, $h_1 \in H_p(X)$, $c_2 \in H^q(Y)$ and $h_2 \in H_q(Y)$.

4.3. LEMMA. If $d = 2$ or 4, cup products of positive dimensional classes in Q_n are zero. If $d = 1$, and $n \geq a \geq b > 0$, then $\bar{1}\{a\} = 0$ and $\{a\}\{b\} = \{a + b - 1\}$.

PROOF. For $d = 2$ or 4, 1.4 shows that Q_n has only odd dimensional cohomology. Since the product of two odd dimensional classes is zero, the lemma follows for $d = 2$ or 4.

When $d = 1$, Q_n is the disjoint union of Q_0 and P^{n-1} by 3.2. Since $\{a\}$ is the generator of $H^{a-1}(P^{n-1};Z_2)$, the formula $\{a\}\{b\} = \{a + b - 1\}$ follows. The unit element in $H^*(Q_n;Z_2)$ is $\bar{1} + \{1\}$, since this assigns the value 1 to each 0-cell in Q_n. Therefore

$$(\bar{1} + \{1\})\{a\} = \{a\} = \{a\}\{1\} .$$

The lemma follows.

4.4. LEMMA. Let $n \geq k \geq 1$. In $H^*(O(n,k);Z_2)$, we have

$$\{a\} \cup \{b_1,\ldots,b_r\} = \{a,b_1,\ldots,b_r\} + \sum_{i=1}^{r}\{b_1,\ldots,b_i + a - 1,\ldots,b_r\}$$

where $a > k$ and $b_i > k$ for all i.

PROOF. The theorem is true for $n = k$, since then all terms in the formula are zero (see 4.1). For $n > k$, it follows by induction on n, using 4.2 a) and 4.3.

Let $\Lambda(n,k)$ be the commutative associative algebra on the generators $\{b\}$ of dimension $b - 1$, where $n \geq b > k$, subject to the relations $\{b\}\{b\} = \{2b - 1\}$ if $2b - 1 \leq n$ and $\{b\}\{b\} = 0$ if $2b-1 > n$.

4.5 THEOREM. Let $n \geq k \geq 1$. Then $H^*(O(n,k);Z_2) \approx \Lambda(n,k)$. If $n \leq m$ and $k \leq \ell$, we have a map $O(n,k) \longrightarrow O(m,\ell)$ which induces a map $H^*(O(m,\ell);Z_2) \longrightarrow H^*(O(n,k);Z_2)$. Under this map, $\{b\} \longrightarrow 0$ if $b > n$ and $\{b\} \longrightarrow \{b\}$ if $b \leq n$.

PROOF. From 3.4, we see that $H^*(O(n,k);Z_2)$ has a vector space basis consisting of the normal classes $\{b_1,\ldots,b_r\}$, with $n > b_1 > \ldots > b_r > k$. 4.4 shows by induction on r, that the normal classes $\{b\}$, with $n \geq k$, generate $H^*(O(n,k);Z_2)$. Also from 4.4, we see that

$$\{b\} \cup \{b\} = \{b,b\} + \{2b-1\} .$$

Referring to our notation 4.1, we see that there is an epimorphism $\Lambda(n,k) \longrightarrow H^*(O(n,k);Z_2)$.

Suppose we have an element $Q \in \Lambda$, whose image in $H^*(O(n,k);Z_2)$ is zero. Q can be expressed as the sum of terms of the form $\{b_1\}\{b_2\}\ldots\{b_r\}$, where $n \geq b_1 > \ldots > b_r > k$. By induction on r, we see from 4.4 that

$$\{b_1\}\{b_2\}\ldots\{b_r\} = \{b_1,\ldots,b_r\} + \text{terms like } \{a_1,\ldots,a_s\}$$

with $s < r$. In Q, if we collect the terms where r is greatest, and apply this formula, we see that $Q = 0$.

4.6. LEMMA. If $d = 2$ or 4, then in $H^*(G(n,k);R)$,

$$(-1)^r\{a\} \cup \{b_1,\ldots,b_r\} = \{a,b_1,\ldots,b_r\} ,$$

where $a > k$, and $b_i > k$ for all i.

PROOF. For $n = k$, the theorem is true since then both sides of the equation are zero. For $n > k$, it follows by induction using 4.2 b) and 4.3.

Let $\Gamma(n,k)$ be the exterior algebra over R, generated by elements $\{b\}$ of dimension $bd - 1$ ($d = 2$ or 4), with $n \geq b > k$.

4.7. THEOREM. $H^*(G(n,k);R) \approx \Gamma(n,k)$ for $d = 2$ or 4. If $n \leq m$ and $k \leq \ell$, we have a map $G(n,k) \longrightarrow G(m,\ell)$ which induces a map $H^*(G(m,\ell);R) \longrightarrow H^*(G(n,k);R)$. Under this map $\{b\} \longrightarrow 0$ if $b > n$ and $\{b\} \longrightarrow \{b\}$ if $b \leq n$.

PROOF. This follows from 4.6 in the same way that 4.5 follows 4.4.

§5. The Pontrjagin Rings of SO(n) and SU(n).

$SO(n)$ and $SU(n)$ are the subgroups of $O(n)$ and $U(n)$ respectively, consisting of matrices with determinant 1. The compositions

$$SO(n) \xrightarrow{i} O(n) \xrightarrow{\pi} O(n,1) \quad \text{and} \quad SU(n) \xrightarrow{i} U(n) \xrightarrow{\pi} U(n,1)$$

are both homeomorphisms. The reason for this is that both in real and in complex n-space, there is exactly one way of completing an (n-1)-frame to an n-frame with determinant 1. We shall identify $SO(n)$ with $O(n,1)$ and $SU(n)$ with $U(n,1)$.

5.1. THEOREM. The Pontrjagin ring $H_*(SO(n);Z_2)$ is the exterior algebra on the normal classes $[b]$ of $H_*(O(n,1);Z_2)$ (see 3.3).

PROOF. The normal cell $(i_1,\ldots,i_r)$ of $O(n)$ consists of matrices of determinant $(-1)^r$ (see 1.4, 1.5 and 3.2). Therefore $SO(n)$, as a subspace of $O(n)$, consists of the normal cells $(i_1,\ldots,i_{2r})$ where $n \geq i_1 > \ldots > i_{2r} > 0$. By 3.5, the normal class of such a cell is $[i_1][1][i_2][1] \ldots [i_{2r}][1]$. Therefore by 3.5, the image of $H_*(SO(n);Z_2) \longrightarrow H_*(O(n);Z_2)$ is the exterior algebra on the elements $[b][1]$ with $b > 1$. Mapping into $H_*(O(n,1);Z_2)$, $[b][1]$ becomes $[b]$, by 3.6.

5.2. THEOREM. Let $[b] \in H_*(U(n,1);R) = H_*(SU(n);R)$ be a normal class $(n \geq b > 1)$. The Pontrjagin ring $H_*(SU(n);R)$ is the exterior algebra over R generated by the classes $[b]$ of dimension $2b - 1$.

PROOF. By 3.6, we know that $H_1(U(n,1);R) = 0$. Therefore $H^1(SU(n);R) = 0$. The composition

$$H^*(U(n,1);R) \xrightarrow{\pi^*} H^*(U(n);R) \xrightarrow{i^*} H^*(SU(n);R)$$

is the identity. From 4.7, we know that $\pi^*\{b\} = \{b\}$ where $n \geq b > 1$. Therefore $i^*\{b\} = \{b\}$ where $n \geq b > 1$. Moreover $H^1(SU(n);R) = 0$, so $i^*\{1\} = 0$. By 4.6 and induction on r, this shows that $i^*\{b_1,\ldots,b_r\} = 0$ if $b_i = 1$ for some i and that otherwise $i^*\{b_1,\ldots,b_r\} = \{b_1,\ldots,b_r\}$. The dual map $i_*\colon H_*(SU(n);R) \longrightarrow H_*(U(n);R)$ therefore satisfies $i_*[b] = [b]$ where $n \geq b > 1$. Since i_* is a monomorphism of Pontrjagin rings, the theorem follows.

We now investigate the embedding $Sp(n) \subset U(2n)$. Let V be quaternionic n-space and let W be complex 2n-space. Let us write every quaternion $q = q_1 + iq_1 + jq_3 + kq_4$, where q_1, q_2, q_3 and q_4 are real, as $(q_1 + iq_2) + j(q_3 - iq_4)$. Then a column vector $(x_1,\ldots,x_n) \in V$ becomes a column vector $(y_1,\ldots,y_{2n}) \in W$, by writing $x_i = jy_{2i-1} + y_{2i}$. This gives an identification of V and W as complex vector spaces. The identification preserves the scalar product of a vector with itself (but not with another vector). Therefore every element of Sp(n) preserves scalar products in W, and we have an embedding $Sp(n) \longrightarrow U(2n)$. We also have maps $Sp(n) \longrightarrow V$ and $U(2n) \longrightarrow W$ obtained by taking the last column of a matrix, or, equivalently, by taking the image of the vector $(0,\ldots,0,1)$ under an element of Sp(n) or U(2n). The diagram

$$\begin{array}{ccc} Sp(n) & \longrightarrow & V \\ \downarrow & & \downarrow \\ U(2n) & \longrightarrow & W \end{array}$$

is commutative. On identifying unit vectors in V and W with S^{4n-1}, we obtain the following commutative diagram

5.3.
$$\begin{array}{ccc} Sp(n) & \longrightarrow & Sp(n,n-1) \\ & & \| \\ \Big\downarrow & & S^{4n-1} \\ & & \| \\ U(2n) & \longrightarrow & U(2n,2n-1) \end{array} .$$

5.4. THEOREM. The embedding $Sp(n) \longrightarrow U(2n)$ induces an epimorphism $H^*(U(2n);R) \longrightarrow H^*(Sp(n);R)$ given by $\{2b\} \longrightarrow \{b\}$ and $\{2b-1\} \longrightarrow 0$, where $n \geq b > 0$. (Recall that $\{b\}$ has dimension $2b-1$ or $4b-1$, according as it denotes a normal class of U(2n) or Sp(n).)

PROOF. The proof is by induction on n. If we take Sp(0) = U(0) to be the identity transformation, the theorem is obviously true.

The following diagram is commutative

$$\begin{array}{ccc} Sp(n-1) & \longrightarrow & U(2n-2) \\ \downarrow & & \downarrow \\ Sp(n) & \longrightarrow & U(2n) \end{array} .$$

Therefore the diagram

$$\begin{array}{ccc} H^*(Sp(n-1) & \longleftarrow & H^*(U(2n-2)) \\ \uparrow & & \uparrow \\ H^*(Sp(n)) & \longleftarrow & H^*(U(2n) \end{array}$$

is commutative. By 4.7, the left hand vertical map sends $\{b\}$ to $\{b\}$ if $0 < b \leq n-1$ and to zero if $b > n-1$. Also by 4.7, the right hand vertical map sends $\{b\}$ to $\{b\}$ if $0 < b \leq 2n-2$ and to zero if $b > 2n-2$. By our induction hypothesis and the commutativity of the diagram, the theorem is true for classes $\{b\} \in H^{2b-1}(U(2n))$, where $b \leq 2n-2$, and we have only to check the theorem for $\{2n-1\}$ and $\{2n\}$. For dimensional reasons, $\{2n-1\} \in H^{4n-3}(U(2n))$ goes into the subalgebra of $H^*(Sp(n))$ generated by elements of the form $\{m\} \in H^{4m-1}(Sp(n))$ where $0 < m < n$. But by 4.7, this subalgebra is sent monomorphically into $H^*(Sp(n-1))$. By the commutativity of the above diagram, we see that $\{2n-1\}$ goes into zero in $H^*(Sp(n))$.

To find the image of $\{2n\} \in H^{4n-1}(u(2n))$ in $H^{4n-1}(Sp(n))$, we use 4.7 and the diagram 5.3. Since both $\{n\} \in H^{4n-1}(Sp(n))$ and $\{2n\} \in H^{4n-1}(U(2n))$ are images of the fundamental class in $H^{4n-1}(S^{4n-1})$, $\{2n\}$ is sent to $\{n\}$ and the theorem is proved.

§6. Cohomology Operations in Stiefel Manifolds.

We can compute cohomology operations in the Stiefel manifolds as follows. From 4.5 and 4.7, we need only know their action in $SO(n) = O(n,1)$, $U(n)$ and $Sp(n)$. We have the monomorphisms

$$\mu^*: \; H^*(O(n,1);Z_2) \longrightarrow H^*(Q_n \times O(n-1,1);Z_2)$$

and

$$\mu^*: \; H^*(U(n);R) \longrightarrow H^*(Q_n \times U(n-1,1);R) \; .$$

By induction on n, we can determine cohomology operations, if we know them in Q_n and their behaviour under cross products. By 3.2 if $d = 2$, Q_n has the homotopy type of $SCP^{n-1} \vee S^1$, so we need only know the operations in CP^{n-1} and their behaviour under cross products (see I 2.1).

To find the operations in Sp(n) we use 5.4 and our knowledge of their action in $U(2n)$.

The only explicit computation of operations which we shall carry out, is the effect of Sq^i on $H^*(O(n,k);Z_2)$ for $k \geq 1$.

Using the notation 4.1, we have

6.1. THEOREM. $Sq^i\{b\} = \binom{b-1}{i}\{b+i\}$ in $H^*(O(n,k);Z_2)$ $(n \geq b > k \geq 1)$. The Cartan formula then gives the action on the other cohomology classes.

PROOF. Under the monomorphism

$$\mu^*: \ H^*(O(n,k);Z_2) \longrightarrow H^*(Q_n \times O(n-1,k);Z_2) ,$$

the image of $\{b\}$ is $\bar{1} \times \{b\} + \{b\} \times \{\ \}$ (see 4.2 a)).

By I 2.4 and 3.2, $Sq^i\{b\} = \binom{b-1}{i}\{b + i - 1\}$ and $Sq^i\bar{1} = 0$ if $i > 0$. The theorem follows.

We shall obtain another description of the $\mathfrak{A}(2)$-module structure in terms of the definitions in II §5.

A <u>stunted projective space</u> P^n_r is the space obtained from the real projective n-space P^n by collapsing the (r-1)-skeleton P^{r-1} to a point. We have a map $P^n \longrightarrow P^n_r$, which induces a monomorphism

$$H^*(P^n_r;Z_2) \longrightarrow H^*(P^n;Z_2) .$$

Let w_s be the non-zero element of $H^s(P^n_r;Z_2)$ for $r \leq s \leq n$. Then, by naturality and I 2.4,

$$Sq^i\ w_s = \binom{s}{i} w_{s+i} \quad \text{if} \quad s + i \leq n$$

and

$$Sq^i\ w_s = 0 \quad \text{if} \quad s + i > n .$$

By 3.2; if $d = 1$, $Q_n/Q_k = P^{n-1}_k$ $(n \geq k \geq 1)$. The map $Q_n \longrightarrow O(n,k)$ induces a map

$$P^{n-1}_k = Q_n/Q_k \longrightarrow O(n,k) .$$

We claim that this map is a homeomorphism into. We prove this claim by induction on n. It is true for $n = k$. Suppose $x,y \in Q_n/Q_k$ have the same image in $O(n,k)$. By our induction hypothesis we can assume $x \in Q_n - Q_{n-1}$. Our claim then follows by 2.3.

By 2.1, a normal cell $(i_1,\ldots,i_r|n,k)$ has dimension greater than 2k if $r \geq 2$. Therefore the 2k-skeleton of $O(n,k)$ is P^{2k}_k if $n > 2k$. If $n \leq 2k$, the n-skeleton of $O(n,k)$ is P^{n-1}_k .

6.2. THEOREM. If $k \geq 1$, then $H^*(O(n,k);Z_2)$ is the free $\mathfrak{A}$-algebra generated by $H^*(P^{n-1}_k;Z_2)$. (See II 5.4 for the definition.)

PROOF. Let us take $\Lambda(n,k)$ to be the same algebra as the one defined just before 4.5. The free $\mathfrak{A}(2)$-algebra on $H^*(P^{n-1}_k;Z_2)$ is isomorphic to $\Lambda(n,k)$ as an algebra, if we let w_b and $\{b + 1\}$ correspond $(n > b \geq k)$. This is because $Sq^b\ w_b = w_{2b}$ if $2b \geq n$ and zero otherwise.

We have only to check that the structure on $\Lambda(n,k)$ as an $\mathfrak{A}(2)$-module, induced by this isomorphism, is the same as the natural structure on $H^*(O(n,k);Z_2)$. In fact by the Cartan formula, we need only check on the generators $\{b\}$. Now

$$Sq^i w_b = \binom{b}{i} w_{b+i} \quad \text{and} \quad Sq^i\{b+1\} = \binom{b}{i}\{b+i+1\}$$

unless $b + i \geq n$, when both equations have zero right hand side. The theorem follows.

§7. Vector Fields on Spheres.

By a vector field on a sphere S^{n-1} we mean a continuous field of tangent vectors, one at each point of S^{n-1}. A set of k vectors on S^{n-1} are <u>independent</u> if, at each point of S^{n-1}, the k vectors are linearly independent. For each positive integer n, let $k(n)$ be the largest integer such that S^{n-1} has $k(n)$ independent vector fields. The complete determination of the function $k(n)$ has been achieved recently by J. F. Adams [3]. Writing n in the form

$$n = 2^{4\alpha+\beta}(2s+1)$$

where α, β and s are integers ≥ 0 and $\beta = 0,1,2,$ or 3, then

$$k(n) = 2^\beta + 8\alpha - 1$$

Thus $k(n) = 0$ if n is odd; and, for small even n, we have

$$n = 2,4,6,8,10,12,14,16,18,20,22,24,26,28,30,32,$$
$$k(n) = 1,3,1,7,\ 1,\ 3,\ 1,\ 8,\ 1,\ 3,\ 1,\ 7,\ 1,\ 3,\ 1,\ 9,$$

The existence of $k(n)$ independent fields was proved by Hurwitz and Radon [5]. The complete proof of these results is beyond the scope of these notes. However we shall establish an upper bound on $k(n)$ which is a step toward the complete result and which gives the least upper bound for $n < 16$.

7.1 THEOREM. (Whitehead and Steenrod [4].) If $n = 2^m(2s+1)$, then $k(n) < 2^m$.

In order to prove this theorem, we first prove a lemma.

7.2. LEMMA. Let $n = 2^m(2s+1)$. If $0 < j < 2^m$, then $\binom{n-j-1}{j} \equiv 0 \mod 2$. Also $\binom{n-2^m-1}{2^m} \equiv 1 \mod 2$ if $s \geq 1$.

PROOF. $n - 1 = 2^m - 1 + s \cdot 2^{m+1}$

$$= 1 + 2^1 + \dots + 2^{m-1} + s \cdot 2^{m+1}.$$

If $j = 2^r + \lambda 2^{r+1} < 2^m$, then the coefficient of 2^r in $(n - 1 - j)$ is zero, while the coefficient of 2^r in j is 1. By I 2.6, $\binom{n-j-1}{j} \equiv 0$ mod 2. If $j = 2^m$, then the coefficient of 2^m in $(n - 2^m - 1)$ is 1. So $\binom{n-2^m-1}{2^m} \equiv 1$

PROOF of 7.1. Given k vectors $v_1, \dots, v_k$, which are linearly independent, we can find an orthonormal basis for the space spanned by $v_1, \dots, v_k$. We simply define by induction $u_1 = v_1$, u_i = projection of v_i onto the space orthogonal to $u_1, \dots, u_{i-1}$. We put $w_i = u_i/|u_i|$. The same formulas enable us to deduce the existence of a field of k-frames from the existence of k linearly independent vector fields on any manifold with a Riemannian metric.

The k-frames tangent to a point of $S^{n-1} \subset R^n$ (R^n is Euclidean n-space), correspond in a one-to-one way with the $(k+1)$-frames at the origin of R^n. (We simply use the last vector to specify the point on S^{n-1}.) The existence of a field of k-frames on an $(n-1)$-sphere is the same as the existence of a cross-section to the fibre bundle

$$O(n,n-k-1) \longrightarrow O(n,n-1) = S^{n-1}$$

(see 1.1). Actually we do not use the fact that this is a fibre bundle.

Suppose that in contradiction to the theorem there are 2^m linearly independent fields on S^{n-1} and $n = 2^m(2s + 1)$. Then $s \geq 1$. There must be a cross-section λ to the fibre bundle

$$\pi: O(n,n-2^m-1) \longrightarrow O(n,n-1) = S^{n-1}.$$

Therefore we must have maps

$$H^*(O(n,n-1);Z_2) \xrightarrow{\pi^*} H^*(O(n,n-2^m-1);Z_2) \xrightarrow{\lambda^*} H^*(O(n,n-1);Z_2)$$

whose composition is the identity. By 4.5 $\pi^*\{n\} = \{n\}$. Therefore $\lambda^*\{n\} = \{n\}$. Now $\{n\}$ is the only non-zero positive dimensional term in $H^*(O(n,n-1);Z_2) = H^*(S^{n-1};Z_2)$. Therefore $\lambda^*\{b\} = 0$ if $n > b$. By 6.1 and 7.2

$$Sq^{2^m}\{n-2^m\} = \binom{n-2^m-1}{2^m}\{n\} = \{n\}.$$

Applying λ^* to both sides we have a contradiction, which proves the theorem.

BIBLIOGRAPHY

[1] C. E. Miller: "The topology of rotation groups," Ann. of Math. 57 (1953) pp. 90-113.

[2] I. Yokota: "On the cellular decompositions of unitary groups," J. Inst. Polytech., Osaka City Univ., 7 (1956) pp. 39-49.

[3] J. F. Adams, "Vector fields on spheres," Ann. of Math. 75 (1962).

[4] N. E. Steenrod and J. H. C. Whitehead, "Vector fields on the n-sphere," Proc. Nat Acad. Sci. U.S.A., 37 (1951), pp. 58-63.

[5] B. Eckmann, "Gruppentheoretischer Beweis des Satzes von Hurwitz-Radon," Comment. Math. Helv., 15 (1942), pp. 358-366.

CHAPTER V.

Equivariant Cohomology.

In §1 we define the equivariant cohomology of a chain complex with a group action and show that the cohomology group is left fixed by inner automorphisms of the group. In §2 we give the basic theorem about the construction of a chain map with a prescribed acyclic carrier, and we define the cohomology groups of a group. In §3 we define a generalized form of the cohomology of a group, in which a topological space also plays a role. In §4 we show that a number of alternative ways of defining products in cohomology groups all lead to the same result. In §5 we find the cohomology of of the cyclic groups and in §6 we consider the restriction map from the cohomology of the symmetric group to the cohomology of the cyclic group. In §7 we use the transfer to obtain more accurate information concerning the restriction map.

§1. Chain Complexes with a Group Action.

1.1. DEFINITIONS. The category of pairs is the category whose objects are pairs (ρ,A), where ρ is a group and A is a left ρ-module. A map $f\colon (\rho,A) \longrightarrow (\pi,B)$ consists of homomorphisms $f_1\colon \rho \longrightarrow \pi$ and $f_2\colon B \longrightarrow A$ such that

$$f_2(f_1(\alpha)b) = \alpha f_2(b)$$

for all $\alpha \in \rho$, $b \in B$. The category of algebraic triples is the category whose objects are triples (ρ,A,K) where ρ and A are as above and K is a chain complex on which ρ acts from the left. A map $f\colon (\rho,A,K) \longrightarrow (\pi,B,L)$ consists of a map $(\rho,A) \longrightarrow (\pi,B)$ in the category of pairs and a chain map $f_\#\colon K \longrightarrow L$ such that $f_\#(\alpha k) = f_1(\alpha)f_\#(k)$ for all $\alpha \in \rho$ and $k \in K$. We say that $f_\#$ and f_2 are equivariant (i.e., commute with the group action).

Let $C^*_\rho(K;A) = \mathrm{Hom}_\rho(K,A)$ be the complex of equivariant cochains on K with values in A. A map $f\colon (\rho,A,K) \longrightarrow (\pi,B,L)$ induces a map

$$f^{\#}\colon\ C^*_\pi(L;B) \longrightarrow C^*_\rho(K;A)$$

via the composition

$$K \xrightarrow{f_\#} L \longrightarrow B \xrightarrow{f_2} A\ .$$

Let $H^*_\rho(K;A)$ be the homology of the complex $C^*_\rho(K;A)$.

1.2. LEMMA. $C^*_\rho(K;A)$ and $H^*_\rho(K;A)$ are contravariant functors from the category of algebraic triples.

1.3. DEFINITION. An <u>automorphism</u> of an algebraic triple (ρ,A,K) is a map $(\rho,A,K) \longrightarrow (\rho,A,K)$ with an inverse. The <u>inner automorphism</u> of (ρ,A,K) determined by $\gamma \in \rho$ is defined by

$$f_1(\alpha) = \gamma\alpha\gamma^{-1},\quad f_2(a) = \gamma^{-1}a,\quad f_\#(k) = \gamma k.$$

If π is a normal subgroup of ρ, then an inner automorphsim of (ρ,A,K) induces an automorphism of (π,A,K).

We repeat all the definitions in 1.3 in the case of a pair (ρ,A), by suppressing all mention of K.

An automorphism of (ρ,A,K) induces an automorphism of $H^*_\rho(K;A)$ by 1.2.

1.4. LEMMA. An inner automorphism of the algebraic triple (ρ,A,K) induces the identity map on $H^*_\rho(K;A)$.

PROOF. The induced map is the identity on the cochain level.

1.5. LEMMA. Let (ρ,A,K) be an algebraic triple. Let π be a normal subgroup of ρ and let $\gamma \in \rho$. Let $g\colon (\pi,A,K) \longrightarrow (\pi,A,K)$ be the automorphism determined by γ. Then the image

$$H^*_\rho(K;A) \longrightarrow H^*_\pi(K;A)$$

is pointwise invariant under the automorphism g^*.

PROOF. Let $f\colon (\rho,A,K) \longrightarrow (\rho,A,K)$ be the inner automorphism determined by γ. Then by 1.2, the following diagram is commutative

$$\begin{array}{ccc} H^*_\rho(K;A) & \xrightarrow{f^*} & H^*_\rho(K;A) \\ \downarrow & & \downarrow \\ H^*_\pi(K;A) & \xrightarrow{g^*} & H^*_\pi(K;A) \end{array}$$

Further, 1.4 shows that $f^* = 1$.

§2. Cohomology of Groups.

A <u>regular</u> cell complex K is a cell complex with the property that the closure of each cell is a finite subcomplex homeomorphic to a closed ball. If K is infinite, we give it the <u>weak topology</u> — that is, a set is open if and only if its intersection with every finite subcomplex is open, (i.e., K is a CW complex). Let K and L be cell complexes. A <u>carrier</u> from K to L is a function C which assigns to each cell $\tau \in K$ a subcomplex $C(\tau)$ of L such that a face of τ is sent to a subcomplex of $C(\tau)$. An <u>acyclic carrier</u> is one such that $C(\tau)$ is acyclic for each $\tau \in K$. Let ρ and π be groups which act on K and L respectively (consistently with their cell structures), and let $h: \rho \longrightarrow \pi$ be a homomorphism. An <u>equivariant carrier</u> is one such that $C(\alpha\tau) = h(\alpha)\, C(\tau)$ for all $\alpha \in \rho$ and $\tau \in K$. Let $\varphi: K \longrightarrow L$ be a chain map: we say φ is <u>carried</u> by C if $\varphi(\tau)$ is a chain in $C(\tau)$ for all $\tau \in K$.

2.1. REMARK. Let K and L be CW complexes. We give $K \times L$ the product cell structure and the CW topology. The chain complex of $K \times L$ is the tensor product of the chain complex of K and the chain complex of L. If K and L are both regular complexes, then $K \times L$ is a regular complex. (According to Dowker [1], the product topology on $K \times L$ defines a space which is homotopy equivalent to the CW complex $K \times L$.)

Let K' be a ρ-subcomplex of a ρ-free cell complex and suppose we have an equivariant chain map $K' \longrightarrow L$. Suppose we have an equivariant acyclic carrier from K to L which carries $\varphi|K'$.

2.2. LEMMA. We can extend φ to an equivariant chain map $\varphi: K \longrightarrow L$ carried by C. If φ_0 and φ_1 are any two such extensions carried by C, then there is an equivariant homotopy $I \otimes K \longrightarrow L$ between φ_0 and φ_1. (ρ acts on $I \otimes K$ by leaving I fixed and acting as before on K.)

PROOF. We arrange a ρ-basis for the cells of $K - K'$ in order of increasing dimension. We must define φ so that $\varphi\partial = \partial\varphi$. Since $C(\tau)$ is acyclic for each τ, we can do this inductively. The second part of the lemma follows from the first, since $I \times K$ is a ρ-free complex (see 2.1), and we can define a carrier from $I \times K$ to L by first projecting onto K and then applying C.

2.3. LEMMA. Given a group π, we can always construct a π-free acyclic simplicial complex W.

PROOF. We give π the discrete topology and form the infinite repeated join

$$W = \pi * \pi * \pi \ldots .$$

This repeated join is a simplicial complex. Taking the join of a complex with a point gives us a contractible space. Any cycle in W must lie in a finite repeated join W'. Such a cycle is homologous to zero in $W' * \pi$. Therefore W is acyclic.

We make π act on W as follows: π acts by left multiplication on each factor π of the join and we extend the action linearly. This action is obvious free and the lemma is proved.

Suppose we have a homomorphism $\pi \longrightarrow \rho$ and W is an acyclic π-free complex and V an acyclic ρ-free complex. Then we have an equivariant acyclic carrier from W to V: for each cell $\tau \in W$, we define $C(\tau) = V$. By 2.2 we can find an equivariant chain map $W \longrightarrow V$, and all such chain maps are equivariantly homotopic.

Therefore a map of pairs $f\colon (\pi,A) \longrightarrow (\rho,B)$ as in 1.1 leads to a map of algebraic triples $(\pi,A,W) \longrightarrow (\rho,B,V)$ which is determined up to equivariant homotopy of the chain map $W \longrightarrow V$. By 1.2 we obtain a well-defined induced homomorphism

$$f^*\colon\ H^*_\rho(V;B) \longrightarrow H^*_\pi(W;A).$$

In the class of π-free acyclic complexes, any two complexes are equivariantly homotopy equivalent, and any two equivariant chain maps going from one such complex to another are equivariantly homotopic. Therefore the groups $H_\pi^*(W;A)$, as W varies over the class, are all isomorphic to each other and the isomorphisms are unique and transitive. We can therefore identify all these cohomology groups and write $H^*(\pi;A)$ instead of $H_\pi^*(W;A)$.

2.4. LEMMA. $H^*(\pi;A)$ is a contravariant functor from the category of pairs (see 1.1).

§3. Proper maps.

Suppose we have a continuous map $f\colon K \longrightarrow L$ between two CW complexes. A carrier C for f is a carrier from K to L such that $f(\tau) \subset C(\tau)$ for all cells $\tau \in K$. The minimal carrier of f is the carrier which assigns to each cell $\tau \in K$ the smallest subcomplex of L containing $f(\tau)$. Every carrier of f contains the minimal carrier. We say f is proper if the minimal carrier is cyclic. If π acts on K, ρ acts on L, $h\colon \pi \longrightarrow \rho$ is a homomorphism and f is equivariant, then the minimal carrier is also equivariant.

3.1. LEMMA. Let K and L be finite regular cell complexes. Let π act on K and ρ act on L, let $h\colon \pi \longrightarrow \rho$ be a homomorphism and let $f\colon K \longrightarrow L$ be a continuous equivariant map. Then f can be factored into proper equivariant maps

$$K \xrightarrow{1} K' \longrightarrow L' \xrightarrow{1} L$$

where K' and L' are barycentric subdivisions of K and L.

PROOF. The first barycentric subdivision of a regular cell complex is a simplicial cell complex, as we see by induction on the dimension. Let L' be the nth barycentric subdivision of L for $n \geq 1$. Let U_i be the open star of the ith vertex x_i of L'. Then $\{U_i\}$ is an open covering of L. We can choose a barycentric subdivision K' of K such that each simplex τ of K' is contained in a set of the form $f^{-1}(U_i)$. Then the minimal carrier of τ consists of simplexes all of which have x_i as a vertex. Therefore $f\colon K' \longrightarrow L'$ is proper. The identity maps $K \longrightarrow K'$ and $L' \longrightarrow L$ are obviously proper. The maps are all equivariant. This proves the lemma.

Note that we were able to choose L' to be any barycentric

subdivision of L. Note also that any subdivision finer than K' would do equally well in the place of K'. Lemma 3.1 is true but more difficult to prove if the words "finite" and "barycentric" are deleted from its statement.

3.2. DEFINITION. The <u>category of geometric triples</u> is defined in the same way as the category of algebraic triples (see 1.1), except that we replace the chain complex K by a finite regular cell complex K and equivariant chain maps $f_{\#}$ by equivariant continuous maps. We say a map of geometric triples $(\pi,A,K) \longrightarrow (\rho,B,L)$ is <u>proper</u> if the continuous map $K \longrightarrow L$ is proper.

Let $f\colon (\pi,A,K) \longrightarrow (\rho,B,L)$ be a map of geometric triples. Let W be a π-free acyclic complex and V a ρ-free acyclic complex (these exist by 2.3). We wish to construct a map

$$f^*\colon\ H^*_\rho(V \times L;B) \longrightarrow H^*_\pi(W \times K;A)$$

where the action of π on $W \times K$ is the <u>diagonal action</u> – that is $\alpha(w,k) = (\alpha w,\alpha k)$ for all $\alpha \in \pi$, $w \in W$ and $k \in K$ – and similarly for the action of ρ on $V \times L$.

If f is proper, let its minimal carrier be C. Then we have the acyclic equivariant carrier from $W \times K$ to $V \times L$ which assigns $V \times C(\tau)$ to any cell of the form $w \times \tau \in W \times K$. By 2.2 this gives us an equivariant chain map $f_{\#}\colon\ W \otimes K \longrightarrow V \otimes L$ which is determined up to equivariant homotopy. If f is not proper we can factorize it into proper maps

$$(\pi,A,K) \longrightarrow (\pi,A,K') \longrightarrow (\rho,B,L') \longrightarrow (\rho,B,L)$$

and define $f_{\#}$ as the composition of three chain maps

$$W \otimes K \longrightarrow W \otimes K' \longrightarrow V \otimes L' \longrightarrow V \otimes L.$$

For a proper map of geometric triples $(\pi,A,K) \longrightarrow (\rho,B,L)$ we now have two different constructions of an equivariant chain map $W \otimes K \longrightarrow V \otimes L$. The first is obtained directly and the second is obtained by factorizing into three maps. The results differ by at most an equivariant homotopy. It is easy to see that the definition of $f_{\#}$ does not depend, up to equivariant homotopy, on the number of times we subdivide K and L in 3.1. It easily follows that if

$$(\pi,A,K) \xrightarrow{f} (\rho,B,L) \xrightarrow{g} (\sigma,C,M)$$

are maps of geometric triples and if U is an acyclic equivariant σ-complex, then $g_{\#}f_{\#}$: $W \otimes K \longrightarrow U \otimes M$ is equivariantly homotopic to $(gf)_{\#}$. Letting $L = I \times K$, it follows that if h,k: $K \longrightarrow M$ are equivariantly homotopic as continuous maps, then $h_{\#}$ and $k_{\#}$ are equivariantly homotopic as chain maps from $W \otimes K$ to $U \otimes M$.

Therefore a map of geometric triples $(\pi,A,K) \longrightarrow (\rho,B,L)$ gives rise to a map of algebraic triples $(\pi,A,W \otimes K) \longrightarrow (\rho,B,V \otimes L)$. As in 2.4 we show that $H^*_\pi(W \otimes K;A)$ does not depend on the choice of W. We have therefore proved

3.3 LEMMA. $H^*_\pi(W \times K;A)$ is a contravariant functor from the category of geometric triples. Induced maps are independent of equivariant homotopies of the variable K.

3.4 If π is a normal subgroup of ρ, A is a ρ-module and K is a finite regular cell complex on which ρ acts, then ρ acts on the geometric triple (π,A,K) by the same formulas as in 1.3. Therefore ρ acts on $H^*_\pi(W \times K;A)$ by 3.3. This action commutes with equivariant maps of the variable K. If $\pi = \rho$, then ρ acts trivially on $H^*_\pi(W \times K;A)$ by 1.4. If K has trivial π-action, then the action of $\gamma \in \rho$ on $H^*_\pi(W \times K;A)$ can be found by extending the automorphism of (π,A) induced by γ (see 1.3) to a map of the algebraic triple (π,A,W) into itself. Using the identity map on K, this gives a map of $(\pi,A,W \otimes K)$ into itself, which induces the automorphism of $H^*_\pi(W \times K;A)$.

If K is a point then $H^*_\pi(W \times K;A)$ is just $H^*(\pi;A)$ and 3.3 reduces to 2.4.

§4. Products.

Let K be a π-free CW complex and L a CW complex on which ρ acts, and suppose we have a homomorphism $\pi \longrightarrow \rho$ and a continuous equivariant map f: $K \longrightarrow L$. By an increasing induction on the dimension of the cells which form a π-basis for K, we can construct an equivariant homotopy $I \times K \longrightarrow L$, which starts by being f and ends as a cellular map. This gives rise to an equivariant chain map $f_{\#}$: $K \longrightarrow L$, which is determined up to equivariant homotopy. We can insist that, during the homotopy, the image of each cell in K stays within the minimal carrier of f. Then $f_{\#}$ is carried by the minimal carrier.

Now f induces a map $g\colon K/\pi \longrightarrow L/\rho$. The map $f_{\#}$ induces a chain map $K/\pi \longrightarrow L/\rho$. This chain map will do for $g_{\#}$ since the equivariant homotopy $I \times K \longrightarrow L$ induces a homotopy $I \times K/\pi \longrightarrow L/\rho$ which starts by being g and ends as a cellular map.

Let K and L be regular cell complexes with group action as above, and let f be proper. Then we can choose $f_{\#}\colon K \longrightarrow L$ in a different way to that above. We can simply apply 2.2 using the minimal carrier of f. However our previous choice of $f_{\#}$ was also carried by the minimal carrier. Therefore the two procedures lead to the same result (up to equivariant homotopy).

Let W be a π-free regular cell complex and let $L = W/\pi$. Let π act on $W \times W$ by the diagonal action. The diagonal $d\colon W \longrightarrow W \times W$ is an equivariant proper map. By the discussion above we have

4.1. LEMMA. Any equivariant diagonal approximation in W induces a chain map $L \longrightarrow L \otimes L$ which is homotopic to a diagonal approximation in L. If W is acyclic, then any equivariant chain map $W \longrightarrow W \otimes W$ will induce a map $L \longrightarrow L \otimes L$ which is homotopic to a diagonal approximation.

Let (π,A,M) and (ρ,B,N) be algebraic triples (see 1.1). Then we have a triple $(\pi \times \rho, A \otimes B, M \otimes N)$. We have a map

$$C^*_\pi(M;A) \otimes C^*_\rho(N;B) \longrightarrow C^*_{\pi\times\rho}(M \otimes N; A \otimes B)$$

defined in an obvious way. This gives us a <u>cross-product</u> or <u>external-product</u> pairing

$$H^*_\pi(M;A) \otimes H^*_\rho(N;B) \longrightarrow H^*_{\pi\times\rho}(M \otimes N; A \otimes B).$$

Let W be an acyclic π-free complex and let V be an acyclic ρ-free complex. Let (π,A,K) and (ρ,B,L) be geometric triples. Then the cross-product above gives us a map

$$H^*_\pi(W \times K;A) \otimes H^*_\rho(V \times L;B) \longrightarrow H^*_{\pi\times\rho}(W \times K \times V \times L; A \otimes B).$$

Now the algebraic triples $(\pi \times \rho,\ A \otimes B,\ W \otimes K \otimes V \otimes L)$ and $(\pi \times \rho,\ A \otimes B,\ W \otimes V \otimes K \otimes L)$ are isomorphic via the map which interchanges

V and K (with a sign change). Here the action of $\pi \times \rho$ on $W \otimes K \otimes V \otimes L$ is given by

$$(\alpha,\beta)(w \otimes k \otimes v \otimes \ell) = (\alpha w \otimes \alpha k \otimes \beta v \otimes \beta \ell)$$

for all $\alpha \in \pi$, $\beta \in \rho$, $w \in W$, $k \in K$, $v \in V$ and $\ell \in L$. The action of $\pi \times \rho$ on $W \otimes V \otimes K \otimes L$ is given by

$$(\alpha,\beta)(w \otimes v \otimes k \otimes \ell) = (\alpha w \otimes \beta v \otimes \alpha k \otimes \beta \ell).$$

Therefore we have an isomorphism between $H^*_{\pi\times\rho}(W \times K \times V \times L;A \otimes B)$ and $H^*_{\pi\times\rho}(W \times V \times K \times L;A \otimes B)$. Since $W \times V$ is a $(\pi\times\rho)$-free acyclic complex, we see that, composing this isomorphism with the cross-product above, we have introduced a cross-product

(4.2) $$H^*_\pi(W \times K;A) \otimes H^*_\rho(V \times L;B) \longrightarrow H^*_{\pi\times\rho}(W \times V \times K \times L;A \otimes B)$$

defined on the functor of 3.3. The image of $u \otimes v$ is denoted by $u \times v$.

We have a diagonal map of geometric triples

$$d\colon (\pi,A \otimes B,K) \longrightarrow (\pi \times \pi,A \otimes B,K \times K),$$

where π acts on $A \otimes B$ in the first triple by the diagonal action. Hence we have a map (see 3.3)

$$d^*\colon H^*_{\pi\times\pi}(W \times W \times K \times K;A \otimes B) \longrightarrow H^*_\pi(W \times K;A \otimes B)$$

where $\pi\times\pi$ acts on $W \times W \times K \times K$ by

$$(\alpha,\beta)(v_1,v_2,k_1,k_2) = (\alpha v_1,\beta v_2,\alpha k_1,\beta k_2)$$

for all $\alpha,\beta \in \pi$, $v_1,v_2 \in W$ and $k_1,k_2 \in K$. Combining d^* with the cross-product of 4.2, we have the cup-product pairing

(4.3) $$H^*_\pi(W \times K;A) \otimes H^*_\pi(W \times K;B) \longrightarrow H^*_\pi(W \times K;A \otimes B).$$

If π is the trivial group, this is the usual cup-product in K. If K is a point, then this is the usual cup-product in the cohomology of a group.

4.4. REMARK. Let W and L be as in 4.1. We can compute cup-products in L by constructing an equivariant diagonal approximation in W. This is particularly useful when L is not a regular cell complex.

§5. The Cyclic Group.

Let W be the unit sphere the space of infinitely many complex variables. That is, every point in W has the form $(z_0, z_1, \ldots, z_r, 0, \ldots)$ where $\Sigma\, \bar{z}_i z_i = 1$. We give W the weak topology. Alternatively W may be described as the CW complex obtained by taking the union of the sequence

$$S^1 \subset S^3 \subset S^5 \subset \ldots.$$

Let n be any integer greater than one. Let $T\colon W \longrightarrow W$ be the transformation defined by

$$T(z_0, z_1, \ldots) = (\lambda z_0, \lambda z_1, \ldots)$$

where $\lambda = e^{2\pi i/n}$. T obviously acts freely and generates a cyclic group π of order n.

We now construct an equivariant cell decomposition for W, which makes W a regular cell complex. We do this in the obvious way for S^1, so as to get n 0-cells $e_0, Te_0, \ldots, T^{n-1}e_0$, and n 1-cells, $e_1, Te_1, \ldots, T^{n-1}e_1$. Let $\partial e_1 = (T-1)e_0$. Now we proceed by induction. $S^{2r+1} = S^{2r-1} * S^1$ (where $*$ means join). Here S^1 can be identified with the set of points $(0, \ldots, 0, z_r, 0, \ldots)$ such that $\bar{z}_r z_r = 1$. We construct a cell decomposition for S^{2r+1} by taking its $(2r-1)$-skeleton to be the cell decomposition for S^{2r-1} already defined by our induction. We let the $2r$-cells of S^{2r+1} be of the form $S^{2r-1} * T^i e_0 = T^i e_{2n}$ and we let the $(2r+1)$-cells be of the form $S^{2r-1} * T^i e_1 = T^i e_{2r+1}$. We then have n cells in each dimension.

Let $N = 1 + T + \ldots + T^{n-1}$ and $\Delta = T - 1$ be elements in the group ring of π. Choosing the orientation of the join correctly, we obtain

$$\partial T^i e_{2r} = N e_{2r-1}$$

and

$$\partial T^i e_{2r+1} = T^i \Delta e_{2r}\,.$$

Therefore the cell complex is π-equivariant and is regular.

Let $\Omega = \Sigma_{0 \leq i \leq j < n}\, T^i \times T^j$ be an element in the group ring $Z(\pi \times \pi)$. $Z(\pi \times \pi)$ acts on $W \otimes W$ in the obvious way. $Z(\pi)$ acts on $W \otimes W$ via the map $Z(\pi) \longrightarrow Z(\pi \times \pi)$ induced by the diagonal $\pi \longrightarrow \pi \times \pi$.

5.1. LEMMA. The equivariant map $d\colon W \longrightarrow W \otimes W$ defined by

$$de_{2i} = \Sigma_{j=0}^{i}\, e_{2j} \otimes e_{2i-2j} + \Sigma_{j=0}^{i-1}\, \Omega\, e_{2j+1} \otimes e_{2i-2j-1}$$

$$de_{2i+1} = \Sigma_{j=0}^{i} \left(e_{2j} \otimes e_{2i-2j+1} + e_{2j+1} \otimes Te_{2i-2j}\right)$$ is a chain map.

PROOF. In $Z(\pi \times \pi)$ we have the relations

$$T\times T - 1\times 1 = 1\times\Delta + \Delta\times T$$
$$(T\times T)\Omega - \Omega = N\times 1 - 1\times N$$
$$1\times 1 + T\times T + \dots + T^{n-1}\times T^{n-1} = 1\times N + \Omega(\Delta\times 1)$$
$$1\times T + T\times T^2 + \dots + T^{n-1}\times 1 = N\times 1 - \Omega(1\times\Delta) .$$

Using these relations the lemma follows by a straightforward calculation.

Let $L = W/\pi$. Since W is contractible and covers L n times, L is an Eilenberg-MacLane space of type $K(Z_n, 1)$. L has one cell, also denoted by e_i, in each dimension. We have $\partial e_{2r} = ne_{2r-1}$ and $\partial e_{2r+1} = 0$ in L. Let w_r be the cochain dual to e_r. Then $H^r(L;Z_n)$ is cyclic of order n and is generated by w_r. Let β: $H^q(L;Z_n) \longrightarrow H^{q+1}(L;Z_n)$ be the Bockstein operator associated with the exact coefficient sequence

$$0 \longrightarrow Z_n \longrightarrow Z_{n^2} \longrightarrow Z_n \longrightarrow 0.$$

5.2. THEOREM. $\beta w_1 = -w_2$; $\beta w_2 = 0$. If n is odd, $w_1^2 = 0$, $w_{2r} = (w_2)^r$ and $w_{2r+1} = (w_2)^r w_1$. If $n = 2$, then $w_r = (w_1)^r$.

PROOF. Since $\partial e_2 = ne_1$,

$$\beta w_1 \cdot e_2 = -(1/n)w_1 \cdot \partial e_2 = -w_1 \cdot e_1 = -1.$$

Therefore $\beta w_1 = -w_2$. Since $\beta^2 = 0$, $\beta w_2 = 0$.

By 4.4 we can compute cup-products in L by using the diagonal of 5.1. In L we therefore have the induced diagonal approximation

$$de_{2i} = \Sigma_{j=0}^{i} e_{2j} \otimes e_{2i-2j} + n(n-1)/2\, \Sigma_{j=0}^{i-1} e_{2j+1} \otimes e_{2i-2j-1}$$
$$de_{2i+1} = \Sigma_{j=0}^{2i+1} e_j \otimes e_{2i-j+1} .$$

The theorem follows.

5.3. COROLLARY. If n is odd, $H^*(L;Z_n)$ is the tensor product of the exterior algebra on w_1 and the polynomial algebra on $\beta w_1 = -w_2$. If $n = 2$, $\beta w_1 = w_2$ and $H^*(L;Z_2)$ is the polynomial algebra on w_1.

§6. The Symmetric Group.

Throughout this section we assume that p is an odd prime. Let $S(p)$ be the symmetric group of permutations of p symbols. We regard

$S(p)$ as acting on the finite field Z_p. Let k be a generator of the multiplicative group of Z_p. Then $k^{p-1} = 1$. Let T be the cyclic permutation $T(i) = i + 1$. It is easy to see that any element of $S(p)$ which commutes with T is a power of T. We define $\gamma \in S(p)$ by $\gamma i = ki$. Then

$$\gamma T \gamma^{-1}(i) = \gamma T(k^{-1}i) = \gamma(k^{-1}i + 1) = i + k = T^k(i).$$

So $\gamma T \gamma^{-1} = T^k$. γ is an odd permutation as we see by letting γ act on $\{0, 1, k, \ldots, k^{p-1}\}$.

Let π be the cyclic group generated by T, and let ρ be its normalizer. Then $\gamma \in \rho$. Moreover, ρ is generated by γ and T. For suppose $\alpha \in \rho$ and $\alpha T \alpha^{-1} = T^j$. Then $j = k^i$ for some i. Therefore

$$\gamma^{-i} \alpha T \alpha^{-1} \gamma^i = \gamma^{-i} T^{k^i} \gamma^i = T.$$

Therefore $\gamma^{-i}\alpha$ commutes with T and is thus a power of T.

Let $Z_p^{(q)}$ be the $S(p)$-module which is Z_p as an abelian group, and with action from $S(p)$ as follows. If q is even, let $Z_p^{(q)}$ be the trivial $S(p)$-module. If q is odd, let $S(p)$ act on $Z_p^{(q)}$ by the sign of the permutation. Now T is an even permutation. Therefore $Z_p^{(q)}$ is a trivial π-module and so if K has trivial π-action

$$H_\pi^*(W \times K; Z_p^{(q)}) = H_\pi^*(W \times K; Z_p) = H^*(W/\pi \times K; Z_p).$$

The following two lemmas will be important in Chapter VII. Let K be a finite regular cell complex with trivial π-action.

6.1. LEMMA. Let q be even, let $r \geq 0$ and let $u \in H^r(K; Z_p)$. Then $w_{2i} \times u \in H_\pi^{2i+r}(W \times K; Z_p^{(q)})$ is invariant under $\gamma \in \rho$ if and only if $i = m(p-1)$ for some m, and $w_{2i-1} \times u \in H_\pi^{2i+r-1}(W \times K; Z_p^{(q)})$ is invariant if and only if $i = m(p-1)$ for some m. (See 3.4 for the definition of the action of γ.)

6.2. LEMMA. Let q be odd, let $r \geq 0$ and let $u \in H^r(K; Z_p)$. Then $w_{2i} \times u \in H_\pi^{2i+r}(W \times K; Z_p^{(q)})$ is invariant under $\gamma \in \rho$, if and only if $i = m(p-1)/2$ for some odd number m, and $w_{2i-1} \times u \in H_\pi^{2i+r-1}(W \times K; Z_p^{(q)})$ is invariant if and only if $i = m(p-1)/2$ for some odd number m.

PROOF. Since γ is an odd permutation, the map $g\colon (\pi, Z_p^{(q)}) \longrightarrow$

$(\pi, Z_p^{(q)})$ induced by γ, is given as follows (see 1.3 and 3.4)

g_2: $Z_p^{(q)} \longrightarrow Z_p^{(q)}$ is -1 if q is odd and $+1$ if q is even;

$g_1(T) = \gamma T \gamma^{-1} = T^k.$

With W as in §5, we must construct $g_\#$: $W \longrightarrow W$ which is g_1-equivariant. Let $g_\# e_{2i} = k^i e_{2i}$ and let $g_\# e_{2i+1} = k^i \Sigma_{j=0}^{k-1} T^j e_{2i+1}$. (In these formulas we regard k as an integer, $1 < k < p$.) We extend $g_\#$ to be a g_1-equivariant map. We easily check that $g_\#$ is a chain map by using the following formulas. Let N and Δ be the elements of $Z(\pi)$ described in §5. Then

$$g_1(N) = N \quad \text{and} \quad g_1(\Delta) = T^k - 1.$$

Let σ be an r-cell of K and let u denote a cochain representative for the class $u \in H^r(K;Z_p)$. Then

$$\begin{aligned} g^\#(w_{2i} \times u)\cdot(e_{2i} \times \sigma) &= g_2[(w_{2i} \cdot g_\# e_{2i})(u \cdot \sigma)] \\ &= g_2[k^i(u \cdot \sigma)] \\ &= \begin{cases} k^i(u \cdot \sigma) & \text{if } q \text{ is even} \\ -k^i(u \cdot \sigma) & \text{if } q \text{ is odd.} \end{cases} \end{aligned}$$

Therefore

$$g^\#(w_{2i} \times u) = \begin{cases} k^i(w_{2i} \times u) & \text{if } q \text{ is even} \\ -k^i(w_{2i} \times u) & \text{if } q \text{ is odd.} \end{cases}$$

Also

$$\begin{aligned} g^\#(w_{2i+1} \times u)\cdot(e_{2i+1} \times \sigma) &= (-1)^r\, g_2[(w_{2i+1} \cdot g_\# e_{2i+1})(u \cdot \sigma)] \\ &= (-1)^r\, g_2[w_{2i+1} \cdot k^i\, \Sigma_{j=0}^{k-1}\, T^j e_{2i+1}](u \cdot \sigma) \\ &= (-1)^r\, g_2(\Sigma_{j=0}^{k-1}\, k^i)(u \cdot \sigma) \\ &= (-1)^r\, g_2(k^{i+1})(u \cdot \sigma) \\ &= \begin{cases} (-1)^r\, k^{i+1} & \text{if } q \text{ is even} \\ (-1)^{r+1}\, k^{i+1} & \text{if } q \text{ is odd.} \end{cases} \end{aligned}$$

Therefore

$$g^\#(w_{2i+1} \times u) = \begin{cases} k^{i+1}(w_{2i+1} \times u) & \text{if } q \text{ is even} \\ -k^{i+1}(w_{2i+1} \times u) & \text{if } q \text{ is odd.} \end{cases}$$

For $w_r \times u$ to be invariant under γ, it is necessary and sufficient that $g^*(w_r \times u) - (w_r \times u) = 0$. The lemmas follow since $k^i = 1$, if and only if $i|p-1$, and any non-zero element of Z_p has an inverse.

§7. The Transfer.

In this section, we shall use the same symbol for a cohomology class and one of its cocycle representatives.

Let π be a subgroup of finite index in ρ. Let K be a ρ-complex and A a ρ-module. Then we have the inclusion

$$i\colon\ C^*_\rho(K;A) \longrightarrow C^*_\pi(K;A)$$

inducing a map

$$i^*\colon\ H^*_\rho(K;A) \longrightarrow H^*_\pi(K;A)$$

We define the <u>transfer</u>

$$\tau\colon\ C^*_\pi(K;A) \longrightarrow C^*_\rho(K;A)$$

as follows: if $u \in C^*_\pi(K;A)$ and $c \in K$, then

$$\tau u \cdot c = \Sigma_{\alpha \in \rho/\pi}\ \alpha u \cdot \alpha^{-1} c,$$

where α ranges over a set of left coset representatives $\{\alpha_i\}$ — that is $\cup_i \alpha_i \pi = \rho$ and $\alpha_i \pi \cap \alpha_j \pi = \emptyset$ if $i \neq j$. We check immediately that the definition of τ is independent of the choice of coset representatives. If $\beta \in \rho$, then

$$\beta^{-1}(\tau u \cdot \beta c) = \Sigma\, \beta^{-1}\alpha_i u \cdot \alpha_i^{-1}\beta c = \tau u$$

since, for any fixed β, the set $\{\beta^{-1}\alpha_i\}$ is a set of left coset representatives for π in ρ. Therefore $\tau u \in C^*_\rho(K;A)$. It is immediate that τ is a chain map. Therefore τ induces a map

$$\tau\colon\ H^*_\pi(K;A) \longrightarrow H^*_\rho(K;A)$$

which is natural for equivariant maps of ρ-complexes K.

Let $[\rho:\pi]$ denote the index of π in ρ — that is, the number of elements in the set $\{\alpha_i\}$.

7.1. LEMMA. The composition

$$C^*_\rho(K;A) \xrightarrow{\ i\ } C^*_\pi(K;A) \xrightarrow{\ \tau\ } C^*_\rho(K;A)$$

is multiplication by $[\rho:\pi]$.

PROOF. If $u \in C^*_\rho(K;A)$, then

$$\tau u \cdot c = \Sigma_i\ \alpha_i u \cdot \alpha_i^{-1} c = \Sigma_i\ u \cdot c = [\rho:\pi]\, u \cdot c.$$

Let σ be a subgroup of ρ. Let z range over a set of representatives of double cosets $\sigma z \pi$ of σ and π in ρ. We write ${}_z\pi = \pi \cap (z^{-1}\sigma z)$ and $\sigma_z = z\pi z^{-1} \cap \sigma$. Let ad_z be the restriction to ${}_z\pi$ of the inner automorphism of ρ induced by z. Then $ad_z\colon {}_z\pi \longrightarrow \sigma_z$ is an isomorphism. We also denote by ad_z the homomorphism

$$C^*_{{}_z\pi}(K;A) \longrightarrow C^*_{\sigma_z}(K;A)$$

given by $ad_z u \cdot c = z(u \cdot z^{-1}c)$ where $c \in K$.

<u>The remainder of this section is not required elsewhere in these notes</u>.

7.2. LEMMA. The following diagram is commutative

$$\begin{array}{ccccc}
C^*_\pi(K;A) & \xrightarrow{\ \tau\ } & C^*_\rho(K;A) & \xrightarrow{\ i\ } & C^*_\sigma(K;A) \\
\Big\downarrow \Sigma i_z & & & & \Big\uparrow \Sigma\tau_z \\
\Sigma_z\, C^*_{{}_z\pi}(K;A) & & \xrightarrow{\quad \Sigma\, ad_z \quad} & & \Sigma_z\, C^*_{\sigma_z}(K;A)
\end{array}$$

PROOF. Let y_z range over a set of left coset representatives of σ_z in σ. By $\cup_y$ and Σ_y we shall mean taking unions or sums over y_z, while keeping z fixed. Now

$$\sigma_z z\pi = z(z^{-1}\sigma_z z)\pi = z({}_z\pi)\pi = z\pi.$$

Therefore

$$\sigma z\pi = \cup_y\, y_z\sigma_z\, z\pi = \cup_y\, y_z\, z\pi$$

and so
$$\rho = \cup_z\, \sigma z\pi = \cup_z \cup_y\, y_z\, z\pi$$

We easily check that the last is a disjoint union. Hence the elements $y_z z$ range over a set of representatives of left cosets of π in ρ.

Suppose $u \in C^*_\pi(K;A)$ and $c \in K$. Then

$$\begin{aligned}
\Sigma_z\, \tau_z\, ad_z(i_z u) \cdot c &= \Sigma_z\, \Sigma_y\, y_z[ad_z(i_z u) \cdot y_z^{-1}c] \\
&= \Sigma_z\, \Sigma_y\, y_z z[i_z u(z^{-1}y_z^{-1}c)] \\
&= \Sigma_{z,y}\, y_z x[u \cdot (y_z z)^{-1}c]
\end{aligned}$$

This proves the lemma.

Now take $\sigma = \pi$. Then $_z\pi = \pi \cap z^{-1}\pi z$ and $\pi_z = z\pi z^{-1} \cap \pi$. Let $m = [\rho:\pi]$.

7.3. LEMMA. If p is a prime not dividing m, then the composition

$$H^*_\rho(K;A) \xrightarrow{i} H^*_\pi(K;A) \xrightarrow{\tau} H^*_\rho(K;A)$$

is an isomorphism of the p-primary part of $H^*_\rho(K;A)$.

PROOF. By 7.1, $\tau i = m$. Also multiplication by m is an isomorphism on the p-primary part of an abelian group.

7.4. LEMMA. Let $u \in H^*_\pi(K;A)$ and suppose $p^s u = 0$. If $\mathrm{ad}_z\, i_{_z\pi} = i_{\pi_z} u$ for all $z \in \rho$, then u is the image under i of some $v \in H^*_\rho(K;A)$ such that $p^s v = 0$. If u is the image of some $v \in H^*_\rho(K;A)$ then $\mathrm{ad}_z\, i_{_z\pi}\, u = i_{\pi_z} u$ for all $z \in \rho$.

PROOF. Suppose that $\mathrm{ad}_z\, i_{_z\pi}\, u = i_{\pi_z} u$ for all $z \in \rho$. We choose m' so that $mm' \equiv 1$ modulo p^s. Then by 7.2

$$i\tau u = \Sigma_z\, \tau_{\pi_z}\, \mathrm{ad}_z\, i_{_z\pi}\, u = \Sigma_z\, \tau_{\pi_z}\, i_{\pi_z}\, u\,,$$

where the sum ranges over a set of representatives of double cosets $\pi z\pi$ in ρ.

From the first paragraph of the proof of 7.2, we see that as y_z runs through a set of left coset representatives of π_z in π, and z runs through a set of representatives of double cosets $\pi z\pi$, the elements $y_z z$ form a set of left coset representatives of π in ρ. Let $m_z = [\pi:\pi_z]$. Then $\Sigma_z m_z = m$. Hence

$$\begin{aligned} i\tau u &= \Sigma_z\, \tau_{\pi_z}\, i_{\pi_z}\, u \\ &= \Sigma_z\, m_z\, u \qquad \text{by 7.1} \\ &= mu\,. \end{aligned}$$

Therefore, on putting $v = \tau m'u$, we obtain the first assertion of the lemma.

The second assertion follows directly from the definitions. This proves the lemma.

7.5. LEMMA. Let π be a normal subgroup of ρ, and let p be prime to $[\rho:\pi]$. Then τi is an isomorphism of the p-primary part of $H^*_\rho(K;A)$. If u is in the p-primary part of $H^*_\pi(K;A)$, it is in the image of $i\colon H^*_\rho(K;A) \longrightarrow H^*_\pi(K;A)$, if and only if $ad_z u = u$ for all $z \in \rho$.

PROOF. This is immediate from 7.3 and 7.4.

7.6. LEMMA. If $m = |\rho|$, then $m\, H^q(\rho;A) = 0$ for $q > 0$.

PROOF. Let K be a ρ-free acyclic complex, and let $\pi = 1$. We apply 7.1 and use the fact that $H^q(K;A) = 0$ if $q > 0$. The lemma follows.

7.7. LEMMA. Let π be a Sylow p-subgroup of ρ (ρ finite). Then $H^*(\pi;A)$ is a p-group in positive dimensions and $i\colon H^*(\rho;A) \longrightarrow H^*(\pi;A)$ maps the p-primary part of $H^*(\rho;A)$ isomorphically onto the subgroup of those elements u such that $ad_z\, i_{z^\pi} u = i_{\pi_z} u$ for each $z \in \rho$.

PROOF. We note that $|\pi| = p^s$ and $m = [\rho:\pi]$ is prime to p. By 7.6, $p^s H^*(\pi;A) = 0$ in positive dimensions. So $H^q(\pi;A)$ is a p-group for $q > 0$. The rest of the lemma follows from 7.4 and 7.3.

7.8. PROPOSITION. Let π be a cyclic group of order p, and let π be a Sylow p-subgroup of ρ. Let σ be the normalizer of π in ρ. Then the monomorphic images of the p-primary parts of $H^*(\rho;A)$ and $H^*(\sigma;A)$ (in positive dimensions) coincide. The image is the subgroup of those elements of $H^*(\pi;A)$ which are invariant under σ.

PROOF. Since $|\pi| = p$, we have

$$\pi \cap z^{-1}\pi z = 1 \quad \text{if} \quad z \notin \sigma \quad \text{and}$$
$$\pi \cap z^{-1}\pi z = \pi \quad \text{if} \quad z \in \sigma.$$

Therefore $i_{\pi_z} = i_{z^\pi} = 0$ in positive dimensions, if $z \notin \sigma$. Therefore by 7.7 the conditions for an element to be in the p-primary part of $\mathrm{Im}(H^*(\rho;A))$ are the same as the conditions for it to be in the p-primary part of $\mathrm{Im}(H^*(\sigma;A))$. If $z \in \sigma$, then

$$ad_z\colon H^*(\pi;A) \longrightarrow H^*(\pi;A)$$

is the automorphism induced by z^{-1} (see 1.3 and the definition of ad_z).

This proves the proposition.

7.9. LEMMA. $H^0(\pi;A)$ is isomorphic to the subgroup of invariant elements of A under π. This isomorphism is natural for maps of (π,A) (see 1.1).

PROOF. This follows immediately from the definition of $H^*(\pi;A)$, since an acyclic π-free complex must be connected.

7.10. COROLLARY. If $\pi \subset \rho$ and A is a ρ-module, then the induced map $H^0(\rho;A) \longrightarrow H^0(\pi;A)$ has an image consisting of those elements of A which are invariant under ρ.

BIBLIOGRAPHY

[1] C. H. Dowker, "Topology of Metric Complexes," Am. Jour. of Math., 74 (1952) pp. 555-577.

[2] H. Cartan and S. Eilenberg, "Homological Algebra," Chapter 12: Finite groups, Princeton University Press (1956).

CHAPTER VI.

Axiomatic Development of the Algebra $\mathfrak{A}(p)$.

In §1 we give the axioms for the P^i. In §2 we define the Steenrod algebra $\mathfrak{A}(p)$ and show it is a Hopf algebra. In §3 we obtain the structure of the dual Hopf algebra. The proofs are very similar to those in the mod 2 case. In §5 we obtain some results about the homotopy groups of spheres and in §6 we derive the Wang sequence .

§1. Axioms.

Let p be an odd prime and let

$$\beta: \ H^q(X;Z_p) \longrightarrow H^{q+1}(X;Z_p)$$

be the Bockstein coboundary operator associated with the exact coefficient sequence

$$0 \longrightarrow Z_p \longrightarrow Z_{p^2} \longrightarrow Z_p \longrightarrow 0.$$

We assume as known that β is natural for mappings of spaces, that $\beta^2 = 0$ and that

$$\beta(xy) = (\beta x)y + (-1)^q x(\beta y) \quad \text{where } q = \dim x.$$

We have the following axioms

1) For all integers $i \geq 0$ and $q \geq 0$ there is a natural transformation of functors which is a homomorphism

$$P^i: \ H^q(X;Z_p) \longrightarrow H^{q+2i(p-1)}(X;Z_p) .$$

2) $P^0 = 1$.

3) If $\dim x = 2k$, then $P^k x = x^p$.

4) If $2k > \dim x$, then $P^k x = 0$.

5) Cartan formula.

$$P^k(xy) = \Sigma_i P^i x \cdot P^{k-i} y .$$

6) Adem relations. If $a < pb$ then

$$P^aP^b = \sum_{t=0}^{[a/p]} (-1)^{a+t} \binom{(p-1)(b-t)-1}{a-pt} P^{a+b-t}P^t .$$

If $a \leq b$ then

$$P^a\beta P^b = \sum_{t=0}^{[a/p]} (-1)^{a+t} \binom{(p-1)(b-t)}{a-pt} \beta P^{a+b-t}P^t$$

$$+ \sum_{t=0}^{[(a-1)/p]} (-1)^{a+t-1} \binom{(p-1)(b-t)-1}{a-pt-1} P^{a+b-t}\beta P^t .$$

We shall prove the axioms in Chapters VII and VIII and we shall show that the other axioms imply Axiom 6). As in Chapter I, we can show that, in the presence of Axiom 1), the Cartan formula above is equivalent to

$$P^k(x \times y) = \Sigma P^ix \times P^{k-i}y .$$

We can also show that P^i commutes with suspension and with

$$\delta: H^i(A;Z_p) \longrightarrow H^{i+1}(X,A;Z_p)$$

as in I 1.2 and I 2.1. Similarly $\beta\delta = -\delta\beta$ and $\beta s = -s\beta$, where s is the suspension.

§2. Definition and Properties of $\mathcal{A}(p)$.

We define the Steenrod algebra $\mathcal{A}(p)$ to be the graded associative algebra generated by the elements P^i of degree $2i(p-1)$ and β of degree 1, subject to $\beta^2 = 0$, the Adem relations and to $P^0 = 1$. A monomial in $\mathcal{A}(p)$ can be written in the form

$$\beta^{\varepsilon_0}P^{s_1}\beta^{\varepsilon_1} \dots P^{s_k}\beta^{\varepsilon_k}$$

where $\varepsilon_i = 0,1$ and $s_i = 1,2,3\dots$. We denote this monomial by P^I, where

$$I = (\varepsilon_0, s_1, \varepsilon_1, s_2, \dots, s_k, \varepsilon_k, 0, 0 \dots).$$

A sequence I is called admissible if $s_i \geq ps_{i+1} + \varepsilon_i$ for each $i \geq 1$. The corresponding P^I, and also P^0, will be called admissible monomials. We define the moment of I to be $\Sigma\, i(s_i + \varepsilon_i)$. Let the degree of I be the degree of P^I, which we denote by $d(I)$.

2.1. PROPOSITION. Each element of $\mathcal{A}(p)$ is a linear combination of admissible monomials.

PROOF. As in I 3.1, we see by a straightforward computation that the Adem relations express any inadmissible monomial as the sum of monomials of smaller moment. The proposition then follows by induction on the moment.

We shall investigate $\mathcal{A}(p)$ by letting it operate on a product of lens spaces. We first prove some lemmas

2.2. LEMMA. Let x and y be mod p cohomology classes in any space such that dim $x = 1$ and dim $y = 2$. Then Axioms 2),3),4) and 5) imply that $P^i x = 0$ unless $i = 0$ and

$$P^i y^k = \binom{k}{i} y^{k+i(p-1)}.$$

PROOF. For $k = 1$, the result follows from §1 Axioms 4),3) and 2). For $k > 1$, it follows by induction on k and the Cartan formula.

2.3. LEMMA. If y is as in 2.2 then Axioms 2),3),4) and 5) imply that $P^i(y^{p^k})$ is y^{p^k} if $i = 0$; zero if $i \neq 0, p^k$; and $y^{p^{k+1}}$ if $i = p^k$.

PROOF. This follows immediately from 2.2 and I 2.6.

Let u be a cohomology class of dimension q. Let I be a sequence of the form $(\varepsilon_0, s_0, \varepsilon_1, s_1, \ldots, s_r, \varepsilon_r, 0 \ldots)$. Then we have the formulas

$$\begin{aligned} \beta(u \times v) &= \beta u \times v + (-1)^q u \times \beta v, \\ P^k(u \times v) &= \Sigma\, P^i u \times P^{k-i} v, \\ P^I(uv) &= \Sigma_{K+J=I}\, (-1)^{q \cdot d(J)}\, P^K u \cdot P^J v, \\ P^I(u \times v) &= \Sigma_{K+J=I}\, (-1)^{q \cdot d(J)}\, P^K u \times P^J v. \end{aligned}$$

Let L and $w_i \in H^i(L;Z_p)$ be as in V §5. Let $X = L \times \ldots \times L = L^{2n}$. Let

$$u_n = y \times x \times y \times x \ldots \ldots \times y \times x \in H^{3n}(L^{2n};Z_p)$$

where $x = w_1$ and $y = -w_2$.

2.4. PROPOSITION. The elements $P^I u_n$ are linearly independent, where I ranges over all admissible sequences $(\varepsilon_0, s_1, \varepsilon_1, \ldots, s_i, \varepsilon_i, 0, \ldots)$ of degree $\leq n$.

PROOF. Let $J_k = (0, p^{k-1}, 0, p^{k-2}, \ldots, 0, p^1, 0, p^0, 0, \ldots)$ and

$$J_k' = (0, p^{k-1}, 0, p^{k-2}, \ldots, 0, p^1, 0, p^0, 1, 0, \ldots).$$

Recall that $\beta x = y$ and $\beta y = 0$ (see V 5.2). Therefore, by 2.3, $P^I x = 0$ unless I is J_k' with a number of pairs of adjacent zeros inserted, or $I = (0,0,\ldots)$: $P^{J_k'}x = y^{p^k}$ and $P^0 x = x$. Also by 2.3, $P^I y = 0$ unless H is J_k with a number of pairs of zeros inserted, or $I = (0,0,\ldots)$: $P^{J_k}y = y^{p^k}$ and $P^0 y = y$. We note that $P^I(x \times y) = 0$ if there is more than one non-zero ε_i in I.

We prove the lemma by induction on n. It is obvious for $n = 1$, since the only monomials of degree ≤ 1 are P^0 and β.

Suppose $\Sigma\, a_I P^I u_n = 0$ $(a_I \in Z_p)$, where the sum is taken over admissible sequences I of a fixed degree q, where $q \leq n$. We wish to prove that each $a_I = 0$. This is done by a decreasing induction on the length $\ell(I)$. Suppose that $a_I = 0$ for $\ell(I) > 2m+1$.

The Künneth theorem asserts that

$$H^{q+3n}(L^{2n}) \approx \Sigma_{s,t}\, H^s(L) \otimes H^t(L) \otimes H^{q+3n-s-t}(L^{2n-2}) .$$

Let g_m be the projection onto the factor with $s = p^m$ and $t = 1$. Let h_m be the projection onto the factor with $s = 2$ and $t = p^m$.

$$(1) \quad P^I u_n = P^I(y \times x \times u_{n-1}) = \Sigma_{J+K+L=I}\, (-1)^{d(L)} P^J y \times P^K x \times P^L u_{n-1} .$$

Let I be admissible. We assert that

(2) if $\ell(I) < 2m+1$, then $h_m P^I u_n = 0$, and

if $\ell(I) = 2m+1$, then $I \geq J_m'$ and

$$h_m P^I u_n = (-1)^i\, y \times y^{p^m} \times P^{I-J_m'}\, u_{n-1},$$

where $i = \deg\,(I - J_m')$. We also assert that

(3) if $\ell(I) < 2m$, then $g_m P^I\, u_n = 0$ and

if $\ell(I) = 2m$, then $I \geq J_m$ and

$$g_m P^I u_n = (-1)^i\, y^{p^m} \times x \times P^{I-J_m}\, u_{n-1},$$

where $i = \deg\,(I - J_m)$.

To prove (2) and (3), we refer to the first paragraph of this proof. We note that a sequence obtained from J_m' by inserting zeros has length greater than $2m+1$, and a sequence obtained from J_m by inserting zeros has length greater than $2m$. Therefore (2) and (3) follow from (1).

We can now apply (2) and (3) to our decreasing induction on $\ell(I)$. Since $a_I = 0$ for $\ell(I) > 2m+1$, we see by applying (2) to our relation that

$$y \times y^{p^m} \times \Sigma_{\ell(I)=2m+1} (-1)^i a_I P^{I-J_m'} u_{n-1} = 0.$$

As I ranges over all admissible sequences of length $(2m+1)$ and degree q, $I-J_m'$ ranges over all admissible sequences of length $\leq 2m$ and degree $q - 2p^m + 1$. By our induction on n, we have $a_I = 0$ when $\ell(I) = 2m+1$.

Now applying (3) to our relation, we see that

$$y^{p^m} \times x \times \Sigma_{\ell(I)=2m} (-1)^i a_I P^{I-J_m} u_{n-1} = 0.$$

As I ranges over all admissible sequences of length $2m$ and degree q, $I - J_m$ ranges over all admissible sequences of length $\leq 2m$ and degree $q - 2p^m + 2$. By our induction on n, we have $a_I = 0$ when $\ell(I) = 2m$.

This completes the proof of the proposition.

Combining 2.1 and 2.4 we obtain

2.5. THEOREM. The admissible monomials form a basis for $\mathfrak{A}(p)$.

2.6. COROLLARY. The mapping $\mathfrak{A}(p) \longrightarrow H^*(L^{2n})$ given by evaluation on u_n, is a monomorphism in degrees $\leq n$.

2.7. THEOREM. Any P^k $(k \neq p^i)$ is decomposable. Therefore $\mathfrak{A}(p)$ is generated by β and P^{p^i} $(i = 0,1,2,\ldots)$.

PROOF. By the Adem relations, P^{a+b} is decomposable if $a < pb$ and $\binom{(p-1)b-1}{a} \not\equiv 0 \bmod p$. Let

$$a + b = k = k_0 + k_1 p^1 + \ldots + k_m p^m$$

where $0 \leq k_i < p$ and $k_m \neq 0$. Let $b = p^m$. Then

$$\begin{aligned}(p-1)b - 1 &= (p^m - 1) + (p-2)p^m \\ &= (p-1)(1 + p^1 + \ldots + p^{m-1}) + (p-2)p^m .\end{aligned}$$

Now

$$a = k - b = k_0 + k_1 p^1 + \ldots + k_{m-1} p^{m-1} + (k_m - 1)p^m$$

So by I 2.6,

$$\binom{(p-1)b-1}{a} \equiv \binom{p-1}{k_0}\binom{p-1}{k_1} \cdots \binom{p-1}{k_{m-1}}\binom{p-2}{k_m - 1} \not\equiv 0.$$

The theorem follows from I 4.1.

2.8. LEMMA. Let X be any space such that $H^*(X;Z_p)$ is a polynomial ring on one generator of dimension $2k$ (possibly truncated by $x^t = 0$

where $t > p$). Then k has the form $k = mp^j$ where m divides $(p-1)$.

PROOF. By §1 Axiom 3), $P^k x = x^p \neq 0$. Therefore by 2.7, $P^{p^i} x \neq 0$ for some $p^i \leq k$. Now $\dim (P^{p^i} x) = 2k + 2p^i(p-1)$. Since $P^{p^i} x = \alpha x^s$ $(\alpha \in Z_p)$ for some integer s, we see that $2k + 2p^i(p-1) = 2ks$. Therefore $p^i(p-1) = k(s-1)$. The lemma follows.

2.9. Theorem. If K is a CW complex with a finite n-skeleton for each n, and $H^*(K;Z)$ is a polynomial ring on one generator of dimension $2k$ (possibly truncated by $x^t = 0$ where $t > 3$), then $k = 1$ or 2.

PROOF. We have a commutative diagram

$$\begin{array}{ccc} C^*(K;Z) \otimes Z & \xrightarrow{\approx} & C^*(K;Z) \\ \downarrow & & \downarrow \\ C^*(K;Z) \otimes Z_p & \underset{\approx}{\xrightarrow{\alpha}} & C^*(K;Z_p) \end{array}$$

where the vertical map on the right is the coefficient homomorphism, and the lower horizontal map makes the diagram commutative.

By the universal coefficient theorem for $C^*(K;Z) \otimes Z_p$, we have an exact sequence

$$0 \longrightarrow H^q(K;Z) \otimes Z_p \xrightarrow{\alpha^*} H^q(K;Z_p) \longrightarrow \mathrm{Tor}(H^{q-1}(K;Z), Z_p) \longrightarrow 0.$$

Since $H^*(K;Z)$ is free, the third term is zero. Therefore using the commutative diagram above, we see that the coefficient homomorphism $H^q(K;Z) \longrightarrow H^q(K;Z_p)$ induces an isomorphism

$$H^q(K;Z) \otimes Z_p \approx H^q(K;Z_p) .$$

Since the coefficient homomorphism is a map of coefficient rings, this isomorphism gives an isomorphism of rings.

Therefore $H^*(K;Z_p)$ is a polynomial ring on one generator x of dimension $2k$ (possibly truncated by $x^t = 0$, where $t > 3$). Since $x^3 \neq 0$, we see from 2.8 for $p = 3$ that $k = m3^i$, where $m = 1$ or 2. Since $x^2 \neq 0$, we see from I 4.5 that $k = 2^j$. Therefore $k = 1$ or 2.

2.10 THEOREM. The map of generators $\psi(P^k) = \Sigma P^i \otimes P^{k-i}$ and $\psi(\beta) = \beta \otimes 1 + 1 \otimes \beta$ extends to a map of algebras

$$\psi \colon \mathfrak{A}(p) \longrightarrow \mathfrak{A}(p) \otimes \mathfrak{A}(p) .$$

PROOF. The proof is the same as that of II 1.1. We merely substitute L^{2n} for the n-fold Cartesian product of infinite dimensional real projective space, and substitute u_n for w.

2.11. THEOREM. $\mathfrak{A}(p)$ is a Hopf algebra with a commutative and associative diagonal map.

PROOF. As in II 1.2.

§3. The Structure of the Dual Algebra.

Let $\mathfrak{A}(p)^*$ be the dual of $\mathfrak{A}(p)$. ($\mathfrak{A}(p)$ is of finite type by 2.5.) Then $\mathfrak{A}(p)^*$ is a commutative associative Hopf algebra with an associative diagonal map. Let ξ_k be the dual of $M_k = P^{J_k}$ and let τ_k be the dual of $M_k' = P^{J_k'}$ in the basis of admissible monomials. (J_k and J_k' are as in 2.4.) Then ξ_k has degree $2(p^k - 1)$ and τ_k has degree $2p^k - 1$. Since τ_k has an odd degree, $\tau_k^2 = 0$.

We define

$$\begin{array}{llll}
\tau(0) = \xi_0 = 1, & \tau(i) = \tau_{i-1} & \text{for } i \geq 1, \\
\xi(i) = \xi_i, & & \text{for } i \geq 0, \\
x(0) = x, & x(i) = y^{p^{i-1}} & \text{for } i \geq 1, \\
y(i) = y^{p^i} & & \text{for } i \geq 0,
\end{array}$$

where x and y are the classes in $H^*(L;Z_p)$ described before 2.4. Let $I = (i_1,\ldots,i_n)$ be a sequence of non-negative integers. We define

$$\begin{aligned}
\tau(I) &= \tau(i_1) \ldots \tau(i_m) \in \mathfrak{A}(p)^* \\
\xi(I) &= \xi(i_1) \ldots \xi(i_m) \in \mathfrak{A}(p)^* \\
x(I) &= x(i_1) \times \ldots \times x(i_m) \in H^*(L^m;Z_p) \\
y(I) &= y(i_1) \times \ldots \times y(i_m) \in H^*(L^m;Z_p)
\end{aligned}$$

Let g(I) be the minimum number of transpositions needed to transfer all zeros in I to the right of I.

The following lemma will enable us to determine the structure of $\mathfrak{A}(p)^*$.

3.1. LEMMA. Let $\alpha \in \mathfrak{A}(p)$. Then

$$\alpha(x_1 \times \ldots \times x_n \times y_1 \times \ldots \times y_m) = \Sigma_{I,J}(-1)^{g(I)} < \tau(I)\xi(J),\alpha > x(I) \times y(J)$$

where the summation ranges over terms where I has length n and J has length m. (The summation is finite since we get a zero contribution unless $\tau(I)\xi(J)$ and α have the same degree.)

PROOF. We prove the formula by induction. It is true for $(n,m) = (0,1)$ or $(1,0)$, since non-zero terms occur only when $\alpha = M_k$ or M'_k by 2.2, 2.3 and V 5.2.

Now suppose the lemma is true for $(0,m-1)$. Let $\psi\alpha = \Sigma_s \alpha'_s \otimes \alpha''_s$. By the Cartan formula

$$\begin{aligned}
\alpha(y_1 \times \dots \times y_m) &= \Sigma_s \alpha'_s y_1 \times \alpha''_s(y_2 \times \dots \times y_m) \\
&= \Sigma_{s,j,J'} < \xi(j),\alpha'_s > < \xi(J'),\alpha''_s > y(J) \\
&\qquad \text{where } J = (j,J') \\
&= \Sigma < \xi(j) \otimes \xi(J'),\alpha'_s \otimes \alpha''_s > y(J) \\
&= \Sigma < \xi(j) \otimes \xi(J'),\psi\alpha > y(J) \\
&= \Sigma < \psi^*(\xi(j) \otimes \xi(J')),\alpha > y(J) \\
&= \Sigma < \xi(J),\alpha > y(J) \quad .
\end{aligned}$$

This proves the lemma for $(0,m)$.

Suppose the lemma has been proved for $(n-1,m)$. By the Cartan formula

$$\begin{aligned}
\alpha(x_1 \times \dots \times x_n \times y_1 \times \dots \times y_m) &= \Sigma_s (-1)^{\deg \alpha''_s} \alpha'_s x_1 \times \alpha''_s(x_2 \times \dots \times x_n \times y_1 \times \dots \times y_m) \\
&= \Sigma_{s,i,I',J} (-1)^{\gamma} < \tau(i),\alpha'_s > < \tau(I')\xi(J),\alpha''_s > x(I) \times y(J) \\
&\qquad \text{where } I = (i,I') \text{ and } \gamma = \deg \alpha''_s + g(I') \\
&= \Sigma_{s,I,J} (-1)^{\delta} < \tau(i) \otimes \tau(I')\xi(J),\alpha'_s \times \alpha''_s > x(I) \times y(J) \\
&\qquad \text{where } \delta = \deg \alpha''_s + g(I') + \deg \alpha'_s (\deg \tau(I')\xi(J)).
\end{aligned}$$

We must compute δ mod 2 for the non-zero terms of the sum. Now, if a term of the sum is non-zero, then $\tau(i)$ and α'_s have the same degree, and $\tau(I')\xi(J)$ and α''_s have the same degree. Since $\xi(J)$ has even degree, we have mod 2

$$\delta \equiv \deg \tau(I') + g(I') + \deg \tau(i) \cdot \deg \tau(I') \quad .$$

Since the number of non-zero terms in I' is congruent mod 2 to deg $\tau(I')$, we see that if $i = 0$,

$$\delta \equiv \deg \tau(I') + g(I') = g(I) .$$

If $i \neq 0$ then $\deg \tau(i) \equiv 1$ and $\delta \equiv g(I') = g(I)$. This proves that the expression above is

$$\Sigma (-1)^{g(I)} < \psi_*(\tau(i) \otimes \tau(I')\xi(J), \alpha > x(I) \times y(J)$$

$$= \Sigma (-1)^{g(I)} < \tau(I)\xi(J), \alpha > x(I) \times y(J) .$$

This proves the lemma.

Let $\mathfrak{A}'$ denote the free, graded, commutative algebra over Z_p generated by $\tau_0, \tau_1, \ldots$ and $\xi_1, \xi_2, \ldots$. As is well known, $\mathfrak{A}'$ is a tensor product

$$E(\tau_0, \tau_1, \ldots) \otimes P(\xi_1, \xi_2, \ldots)$$

of an exterior algebra and a polynomial algebra (recall that τ_i has odd degree and so $\tau_i^2 = 0$). Since $\mathfrak{A}'$ is free and $\mathfrak{A}(p)^*$ is commutative, the map of the generators of $\mathfrak{A}'$ into $\mathfrak{A}(p)^*$ extends in just one way to a homomorphism of algebras $\mathfrak{A}' \longrightarrow \mathfrak{A}(p)^*$.

3.2. THEOREM. The map $\mathfrak{A}' \longrightarrow \mathfrak{A}(p)^*$ is an isomorphism.

PROOF. We first show that $\mathfrak{A}' \longrightarrow \mathfrak{A}(p)^*$ is an epimorphism. Suppose $< \tau(I)\xi(J), \alpha > = 0$ for all choices of I and J. By 3.1,

$$\alpha(x_1 \times \ldots \times x_n \times y_1 \times \ldots \times y_m) = 0$$

for all choices of m and n. But, 2.6 shows that in this case $\alpha = 0$. Therefore $\mathfrak{A}' \longrightarrow \mathfrak{A}(p)^*$ is an epimorphism.

We now show that the map $\mathfrak{A}' \longrightarrow \mathfrak{A}(p)^*$ is an isomorphism, by showing that in each dimension, the ranks of $\mathfrak{A}'$ and $\mathfrak{A}(p)^*$ as vector spaces over Z_p are the same. We have only to show that the ranks of $\mathfrak{A}'$ and $\mathfrak{A}(p)$ are the same in each degree.

We write $\xi^I = \tau_0^{\varepsilon_0} \xi_1^{r_1} \tau_1^{\varepsilon_1} \ldots \xi_k^{r_k} \tau_k^{\varepsilon_k}$, where $I = (\varepsilon_0, r_1, \varepsilon_1, \ldots, r_k, \varepsilon_k, 0, \ldots)$ and $\varepsilon_i = 0$ or 1, $r_i \geq 0$. The monomials ξ^I, which form a basis of $\mathfrak{A}'$, correspond in a one-to-one way with such sequences I. The admissible monomials $P^{I'} \in \mathfrak{A}$ correspond to sequences of integers $I' = (\varepsilon_0', s_1, \ldots, s_k, \varepsilon_k', 0, \ldots)$ where $s_i \geq ps_{i+1} + \varepsilon_i'$ for each i, and $\varepsilon_i = 0$ or 1. It remains to set up a one-to-one correspondence between the sequences I and I', preserving the degrees of the corresponding monomials.

Let R_k be the sequence with zeros everywhere except for 1 in the $2k$-th place. Let Q_k be the sequence with zeros everywhere except in the $(2k+1)$-th place. Let

$$R_k' = (0,p^{k-1},0,p^{k-2},\ldots,0,p^1,0,p^0,0,\ldots)$$
$$Q_k' = (0,p^{k-1},0,p^{k-2},\ldots,0,p^1,0,p^0,1,0,\ldots) .$$

The map from sequences I to sequences I' can now be defined by extending the map already defined on R_k and Q_k to be additive (with respect to coordinates). Then if

$$I = (\varepsilon_0,r_1,\varepsilon_1,\ldots,r_k,\varepsilon_k,0,\ldots) \longrightarrow I' = (\varepsilon_0',s_1,\varepsilon_1',\ldots,s_k,\varepsilon_k',0,\ldots) ,$$

we have $\varepsilon_i' = \varepsilon_i$ and

$$s_i = (r_i + \varepsilon_i) + (r_{i+1} + \varepsilon_{i+1})p^1 + \ldots + (r_k + \varepsilon_k)p^{k-1} .$$

Solving for r_i in terms of s_i, we see that

$$r_i + \varepsilon_i = s_i - ps_{i+1} .$$

Therefore, given an admissible sequence I', we obtain a unique sequence I with $\varepsilon_i = 0$ or 1 and $r_i \geq 0$, and vice versa. A computation of degrees shows that

$$\deg \xi^I = \deg P^{I'} = \Sigma_1^k r_j 2(p^j - 1) + \Sigma_0^k \varepsilon_j(2p^j - 1).$$

This completes the proof of the theorem.

3.3. THEOREM. The diagonal map $\varphi^*: \mathfrak{A}^* \longrightarrow \mathfrak{A}^* \otimes \mathfrak{A}^*$ is given by

$$\varphi^*\xi_k = \Sigma_{i=0}^k \xi_{k-i}^{p^i} \otimes \xi_i \quad \text{and}$$
$$\varphi^*\tau_k = \tau_k \otimes 1 + \Sigma_{i=0}^k \xi_{k-i}^{p^i} \otimes \tau_i .$$

PROOF. Let $\alpha,\beta \in \mathfrak{A}$. We have to show that

$$< \varphi^*\xi_k,\alpha \otimes \beta > = \Sigma < \xi_{k-i}^{p^i} \otimes \xi^i,\alpha \otimes \beta > \quad \text{and}$$
$$< \varphi^*\tau_k,\alpha \otimes \beta > = < \tau_k \otimes 1,\alpha \otimes \beta > + < \Sigma\, \xi_{k-i}^{p^i} \otimes \tau_i,\alpha \otimes \beta > .$$

That is, we have to show

$$(1) \quad < \xi_k,\alpha\beta > = \Sigma < \xi_{k-i}^{p^i},\alpha > < \xi_i,\beta >, \quad \text{and}$$

(2) $\quad < \tau_k, \alpha\beta > \ = \ < \tau_k, \alpha > < \xi_0, \beta > \ + \ \Sigma < \xi_{k-i}^{p^i}, \alpha > < \tau_i, \beta >$.

Let x and y have the same meaning as in 3.1. In the same way as in II 2.3, we prove that

$$\alpha y^{p^i} \ = \ \Sigma_a < \xi_a^{p^i}, \alpha > y^{p^{a+i}} \ .$$

Now

$$\Sigma_k < \xi_k, \alpha\beta > y^{p^k} \ = \ \alpha\beta y \qquad\qquad\qquad \text{by 3.1}$$

$$= \ \alpha \ \Sigma_i < \xi_i, \beta > y^{p^i}$$

$$= \ \Sigma_{a,i} < \xi_a^{p^i}, \alpha > < \xi_i, \beta > y^{p^{i+a}} \ .$$

Equating coefficients of powers of y, we see that (1) holds. It remains to prove (2). Now

$$< \xi_0, \alpha\beta > x \ + \ \Sigma_k < \tau_k, \alpha\beta > y^{p^k} \ = \ \alpha\beta x \qquad\qquad \text{by 3.1}$$

$$= \ \alpha[< \xi_0, \beta > x \ + \ \Sigma_i < \tau_i, \beta > y^{p^i}]$$

$$= \ < \xi_0, \beta > < \xi_0, \alpha > x \ + \ \Sigma_a < \xi_0, \beta > < \tau_a, \alpha > y^{p^a}$$

$$+ \ \Sigma_{i,a} < \tau_i, \beta > < \xi_a^{p^i}, \alpha > y^{p^{a+i}} \ .$$

Equating coefficients of powers of y, we obtain (2).

This proves the theorem.

§4. Ideals.

Let M_k be the ideal of $\mathcal{A}^*$ generated by

$$\xi_1^{p^k}, \ \xi_2^{p^{k-1}}, \dots, \xi_k^p, \xi_{k+1}, \tau_{k+1}, \dots, \xi_{k+i}, \tau_{k+i}, \dots \ .$$

Then M_k is a Hopf ideal by 3.3. Therefore $\mathcal{A}^*/M_k$ is a __finite__ Hopf algebra. Its dual is a Hopf subalgebra $\mathcal{A}_k \subset \mathcal{A}$. Arguing as in II 3.2 (with minor embellishments), we see that $\beta, P^1, \dots, P^{p^{k-1}}$ are all elements of $\mathcal{A}_k$. It follows that

4.1 THEOREM. $\mathcal{A}$ is the union of the sequence $\mathcal{A}_k$ of finite Hopf subalgebras.

If A is any commutative algebra over Z_p and $\lambda: A \longrightarrow A$ is defined by $\lambda x = x^p$, then λ is a map of algebras. Moreover λ commutes with maps of algebras. Hence if A is a Hopf algebra, λ is a map of Hopf algebras.

Then $\lambda\colon \mathfrak{A}(p)^* \longrightarrow \mathfrak{A}(p)^*$ multiplies degrees by p. The kernel of λ is the ideal generated by $\tau_0, \tau_1, \ldots$.

4.2. LEMMA. If $x \in \mathfrak{A}^*$ and $P^I \in \mathfrak{A}$, then $x^p \cdot P^I = x \cdot P^J$ if $I = pJ$, and $x^p \cdot P^I = 0$ otherwise. (Notice that if $I = pJ$, neither P^I nor P^J can contain β as a factor.)

PROOF. Without loss of generality, we can suppose x is a monomial in $\xi_1, \xi_2, \ldots$ and $\tau_0, \tau_1, \ldots$. Let $x^p \cdot P^I \neq 0$: then x can contain no factor of the form τ_i, since $\tau_i^{\,2} = 0$. Therefore x has even dimension. We have

$$\begin{aligned} x^p \cdot P^I &= \psi_*(x \otimes \ldots \otimes x) \cdot P^I \\ &= (x \otimes \ldots \otimes x) \cdot \psi P^I \\ &= \Sigma\, (x \otimes \ldots \otimes x)(P^{J_1} \otimes \ldots \otimes P^{J_p}) \end{aligned}$$

where the summation is over all sequences $J_1, \ldots, J_p$ such that $J_1 + \ldots + J_p = I$. So

$$x^p .\, P^I = \Sigma\, (x\, P^{J_1}) \ldots (x\, P^{J_p}) .$$

If, in some term of the sum, two of the J_i's are not equal, then cyclic permutation gives p equal terms of the sum. These cancel out mod p. So, if $x^p P^I \neq 0$, $I = pJ$ and $x^p P^I = (x\, P^J)^p = x\, P^J$. This proves the lemma.

Let $\mathfrak{A}'$ be the Hopf subalgebra of $\mathfrak{A}(p)$ generated by P^j $(j = 1, 2 \ldots)$. Let $\lambda^*\colon \mathfrak{A}(p) \longrightarrow \mathfrak{A}(p)$ be the map dual to λ.

4.3. PROPOSITION. λ^* is a map of Hopf algebras, which divides degrees by p. The image of λ^* is $\mathfrak{A}'$, and its kernel is the ideal generated by P^1 and β.

$$\lambda^* P^I = P^J \quad \text{if } I = pJ$$

$$\lambda^* P^I = 0 \quad \text{otherwise} .$$

PROOF. Using 4.2, we see that we have only to check that the kernel of λ^* is contained in the ideal generated by β and P^1. Applying the formulas for λ^* to a linear combination of admissible monomials, we see that we have only to prove that P^k is in the ideal generated by P^1, if k is not a multiple of p. By the Adem relations

$$P^1P^b = (-1)\binom{(p-1)b-1}{1} P^{b+1} = (b+1)P^{b+1} .$$

Therefore P^k is in the ideal generated by P^1 if k is not a multiple of p.

This proposition has been used by Wall [2] and Novikov [1].

4.4. PROPOSITION. If we abelianize $\mathfrak{A}(p)$, we obtain a Hopf algebra, which is the tensor product of $E(\beta)$, the exterior algebra on β, and the divided polynomial ring on $P^1, P^2 \ldots$, i.e.,

$$P^hP^k \equiv \binom{h+k}{h} P^{h+k} \qquad [\mathfrak{A}, \mathfrak{A}].$$

PROOF. Let I be the ideal generated by all commutators in $\mathfrak{A}(p)$. Let $A = \mathfrak{A}/I$. Then A and $A \otimes A$ are commutative algebras. Consider the composition

$$\mathfrak{A} \xrightarrow{\psi} \mathfrak{A} \otimes \mathfrak{A} \longrightarrow A \otimes A .$$

This is an algebra homomorphism into a commutative algebra and is therefore zero on I. Therefore

$$\psi(I) \subset \mathfrak{A} \otimes I + I \otimes \mathfrak{A}$$

and I is a Hopf ideal. Therefore A is a Hopf algebra. A^* consists of all elements $x \in \mathfrak{A}^*$ such that ψx is symmetric. Therefore $\tau_0, \xi_1 \in A^*$.

Suppose that $\Sigma\, a_J \xi^J \in A^*$ $(a_J \in Z_p)$. Then $\Sigma\, a_J \varphi^* \xi^J$ is symmetric. Let $J = (\varepsilon_0, r_1, \varepsilon_1, \ldots, r_k, \varepsilon_k, 0, \ldots)$. We collect terms in $\varphi^* \xi^J$ of the form $\xi_1{}^n \otimes \xi'$ and $\xi'' \otimes \xi_1{}^m$, where m and n are maximal. A short calculation shows that these terms are

$$\xi_1{}^n \otimes \tau_0{}^{\varepsilon_1} \xi_1{}^{r_2} \tau_1{}^{\varepsilon_2} \cdots \xi_{k-1}{}^{r_k} \tau_{k-1}{}^{\varepsilon_k} \quad \text{where } n = \Sigma_1^k (r_i + \varepsilon_i)p^{i-1}$$

$$\text{and} \quad \xi_1{}^{pr_2} \cdots \xi_{k-1}{}^{pr_k} \otimes \xi_1{}^m \quad \text{where } m = \Sigma_1^k r_i .$$

We note that $\Sigma\,(r_i + \varepsilon_i)p^{i-1} \geq \Sigma\, r_i$, and that we have equality only if $\varepsilon_i = 0$ for $i \geq 1$ and $r_i = 0$ for $i \geq 2$.

In $\Sigma a_J \xi^J$ we select those terms for which $\Sigma_1^k (r_i + \varepsilon_i)p^{i-1}$ is maximal. By symmetry, we must have $\varepsilon_i = 0$ for $i \geq 1$ and $r_i = 0$ for $i \geq 2$. Such terms are in the algebra generated by τ_0 and ξ_1. An induction on $\Sigma_1^k (r_i + \varepsilon_i)p^{i-1}$ therefore shows that A^* is the subaglebra generated by τ_0 and ξ_1.

Dualizing, we see that A has the structure described in the proposition.

§5. Homotopy Groups of Spheres.

If G is an abelian group, we let G_p be the subgroup of elements whose orders are powers of the prime p. If G is finitely generated then G can be expressed as the direct sum

$$G = F + \Sigma_p G_p$$

where F is a free group. In this case we can talk of the p-primary part of an element of G, by which we mean the component in G_p.

5.1. THEOREM. $\pi_i(S^3)$ is finite for $i > 3$.

$$\pi_i(S^3)_p = \begin{cases} 0 & \text{if } i < 2p , \\ Z_p & \text{if } i = 2p . \end{cases}$$

Let $f\colon S^{2p} \longrightarrow S^3$ represent an element of $\pi_{2p}(S^3)$ with a non-zero p-primary part and let E be a $(2p+1)$-cell. Then, if $L = S^3 \cup_f E$,

$$P^1\colon H^3(L;Z_p) \longrightarrow H^{2p+1}(L;Z_p)$$

is an isomorphism. (When $p = 2$, replace P^1 by Sq^2, see I 2.3).

5.2. COROLLARY. Let $g\colon S^{n+2p} \longrightarrow S^{n+3}$ be the n-fold suspension of f, and let $M = S^n L$. Then

$$P^1\colon H^{n+3}(M;Z_p) \longrightarrow H^{n+2p+1}(M;Z_p)$$

is an isomorphism. Therefore

$$\pi_{n+2p}(S^{n+3})_p \neq 0.$$

PROOF of 5.2. As P^1 commutes with suspension, the first part follows. Since M is formed by attaching a $(2p+n+1)$-cell to S^{n+3} with the map g, the second part follows by taking f to be the generator of $\pi_{2p}(S^3)_p$.

In fact the following stronger result can be deduced from 5.1 by using [3] Chapter XI, Theorem 8.3 and Corollary 13.3.

5.3. COROLLARY. If p is an odd prime, then

$$\pi_{n+i}(S^{n+3})_p = \begin{cases} 0 & \text{if } i < 2p , \\ Z_p & \text{if } i = 2p . \end{cases}$$

The remainder of this section will be concerned with the proof of 5.1. We shall rely heavily on Serre's mod $\mathcal{C}$ theory. We refer the reader to [3] Chapter X or to [4].

We would like to compute the homotopy groups of S^3 by applying the (mod $\mathcal{C}$) Hurewicz theorem. But the Hurewicz theorem in dimension n only applies to spaces that are $(n-1)$-connected (mod $\mathcal{C}$). So $\pi_3(S^3) \approx Z$ is an obstacle to this program. We therefore construct a space X which has the same homotopy groups as S^3 except that $\pi_3(X) = 0$, and then apply the Hurewicz theorem to X. The definition of X, which is rather long, follows.

5.4. DEFINITION. Let π be an abelian group and let $n \geq 2$ be an integer. $K(\pi,n)$ will denote any space whose homotopy groups are all zero except for π_n which is isomorphic to π. Such a space is called an Eilenberg-MacLane space.

5.5 THEOREM. For any abelian group π and any integer $n \geq 2$, there exists a CW-complex which is a $K(\pi,n)$.

REMARK. We can easily show by obstruction theory that all such CW-complexes are homotopy equivalent.

PROOF. Let π be generated by elements x_i with relations r_j between the x_i's. We take a bouquet of n-spheres, one for each x_i. For each relation

$$r_j = \alpha_{1j}\, x_1 + \alpha_{2j}\, x_2 + \ldots + \alpha_{mj}\, x_m$$

where each α is an integer, we map an n-sphere into the bouquet with degree α_{1j} on the n-sphere corresponding to x_1, with degree α_{2j} on the n-sphere corresponding to x_2 and so on. We attach one $(n+1)$-cell to the bouquet for each relation r_j with this map. We now kill successively the homotopy groups in dimensions $n+1$, $n+2$, etc., by attaching cells of dimensions $n+2$, $n+3$, etc.

Computing the nth homology group by using the cell structure, we see that $H_n \approx \pi$. By the Hurewicz theorem we have constructed a $K(\pi,n)$.

5.6. If K is a path-connected topological space with base-point x, let PK be the paths in K starting at x and let ΩK be the loops in K based on x. We have the <u>standard fibration</u> $p\colon PK \longrightarrow K$ obtained by sending a path in K to its end-point. The fibre is ΩK. (See [3] Chapter III.) Note that PK is contractible. By the homotopy exact sequence for a fibration, $\partial\colon \pi_i(K) \approx \pi_{i-1}(\Omega K)$. If K is an Eilenberg-MacLane space of type $K(\pi,n)$, then this shows that ΩK is an Eilenberg-MacLane space of type $K(\pi,n-1)$.

Let $K = K(Z,3)$ be a CW-complex. Let $S^3 \longrightarrow K$ be a map which represents a generator of $\pi_3(K) \approx Z$. Let $p\colon X \longrightarrow S^3$ be the fibration induced by the standard fibration over K. We have the commutative diagram

$$\begin{array}{ccc} X & \longrightarrow & PK \\ \downarrow p & & \downarrow p \\ S^3 & \longrightarrow & K \end{array}$$

where the vertical maps are fibrations with fibre ΩK, which is a $K(Z,2)$. By the homotopy exact sequences of the fibrations, $\pi_3(X) = 0$ and $p^*\colon \pi_i(X) \approx \pi_i(S^3)$ for $i \neq 3$.

We now find $H_*(X)$ and apply the Hurewicz theorem (mod $\mathcal{C}$) to X to find the first non-vanishing higher homotopy group (mod $\mathcal{C}$) of S^3. The usual method of finding the homology of a fibre space is by using a spectral sequence. In this simple case (base space a sphere), the spectral sequence reduces to the Wang sequence.

5.7. THEOREM. Let $X \longrightarrow S^n$ be a fibration with fibre F. Then we have an exact sequence (Wang's sequence)

$$H^i(X;A) \longrightarrow H^i(F;A) \xrightarrow{\theta} H^{i-n+1}(F;A) \longrightarrow H^{i+1}(X;A)$$

where A is a commutative ring with a unit. Moreover, θ is a derivation: that is, if $x \in H^i(F;A)$, and $y \in H^j(F;A)$, then

$$\theta(xy) = \theta x\cdot y + (-1)^{(n-1)i}\, x\cdot\theta y.$$

PROOF. We refer the reader to [5] p. 471 for a proof by spectral sequences or to the next section of this chapter for a proof not using spectral sequences.

5.8. LEMMA. If $k > 0$, then $H_{2k}(X;Z) = Z_k$ and $H_{2k-1}(X;Z) = 0$.

PROOF. We have the fibration $X \longrightarrow S^3$ with fibre ΩK which is a $K(Z,2)$. Now complex projective space of infinite dimension is also a $K(Z,2)$ and therefore $H^*(\Omega K)$ is a polynomial ring on a two-dimensional generator u. We have the exact sequence (see 5.7)

$$H^i(X;Z) \longrightarrow H^i(\Omega K;Z) \xrightarrow{\theta} H^{i-2}(\Omega K;Z) \longrightarrow H^{i+1}(X;Z) .$$

In order to find $H^*(X;Z)$, we need only find the derivation θ.

Now since X is 3-connected, $H^i(X) = 0$ for $i \leq 3$. Hence $\theta u = \pm 1$. Changing the sign of u, we can ensure that $\theta u = 1$. Since θ is a derivation, $\theta u^n = n\, u^{n-1}$ by induction on n. Therefore

$$H^{2k}(X;Z) = 0 \quad\text{and}\quad H^{2k+1}(X;Z) = Z_k .$$

Let us first consider the class $\mathcal{C}$ of abelian groups which are finitely generated. By the (mod $\mathcal{C}$) Hurewicz Theorem, the homotopy groups of simply connected finite complexes are finitely generated. Therefore $\pi_i(S^3)$ is finitely generated for all i, and so $\pi_i(X)$ is finitely generated for all i. Hence $H_i(X;Z)$ is finitely generated for all i. By the universal coefficient theorem we deduce the lemma.

We now take the class $\mathcal{C}$ of finite abelian groups and deduce from the lemma that $\pi_i(X) \approx \pi_i(S^3)$ is finite for all $i > 3$.

Taking the class $\mathcal{C}$ to consist of all finite abelian groups with orders prime to p, we deduce that

$$\pi_i(S^3)_p = \begin{cases} 0 & \text{if } i < 2p , \\ Z_p & \text{if } i = 2p . \end{cases}$$

This proves the first part of 5.1.

Let $f\colon S^{2p} \longrightarrow S^3$ be as in the statement of 5.1. Let L be S^3 with a $(2p+1)$-cell adjoined with the map f. We can extend the map

$S^3 \longrightarrow K(Z,3)$ which we have been using to a map $L \longrightarrow K(Z,3)$ since $\pi_{2p}(K(Z,3)) = 0$. Let $Y \longrightarrow L$ be the fibration induced by the standard fibration over $K(Z,3)$. By the cell structure of L, we have

$$\pi_i(L,S^3) = 0 \quad \text{for } i < 2p+1 \text{ and}$$

$$\pi_{2p+1}(L,S^3) = Z.$$

Moreover the boundary map

$$\pi_{2p+1}(L,S^3) \longrightarrow \pi_{2p}(S^3)$$

maps the generator of the group on the left onto the element of $\pi_{2p}(S^3)$ represented by f. By the homotopy exact sequence for (L,S^3) we deduce that

$$\pi_i(L) \approx \pi_i(S^3) \quad \text{for } i < 2p, \quad \pi_{2p}(L)_p = 0.$$

By the same reasoning which gave us the homotopy groups of X in terms of those of S^3, we find that

$$\pi_i(Y) = 0 \text{ for } i < 4, \quad \pi_i(S^3) \approx \pi_i(Y) \text{ for } 4 \leq i < 2p,$$

$$\pi_{2p}(Y)_p = 0 .$$

By the (mod) Hurewicz Theorem, $H^i(Y;Z_p) = 0$ for $0 < i \leq 2p$.

5.9. DEFINITION. Suppose we have a fibration $p: E \longrightarrow B$ with fiber F over $b \in B$. Then we have the maps

$$H^n(B,b) \xrightarrow{\rho^*} H^n(E,F) \xleftarrow{\delta} H^{n-1}(F).$$

We say $x \in H^{n-1}(F)$ is <u>transgressive</u> if $\delta x \in \operatorname{Im} p^*$. If our coefficients are Z_p, then a transgressive element is mapped into a transgressive element by any element of the Steenrod algebra mod p.

We show that in the fibration $Y \longrightarrow L$ with fibre ΩK, the generating class $u \in H^2(\Omega K;Z_p)$ is transgressive. From the exact sequences for the pair $(Y,\Omega K)$ and since Y is 3-connected we have the diagram

$$\begin{array}{ccccc} \pi_3(L,x) & \xleftarrow[\approx]{\rho_*} & \pi_3(Y,\Omega K) & \xrightarrow{\approx} & \pi_2(\Omega K) \approx Z \\ \downarrow\approx & & \downarrow & & \downarrow\approx \\ H_3(L,x) & \xleftarrow{\rho_*} & H_3(Y,\Omega K;Z) & \xrightarrow{\approx} & H_2(\Omega k;Z) \end{array}$$

where the vertical maps are Hurewicz homomorphisms and x is the base-point in L. By the universal coefficient theorem, $u \in H^2(\Omega K;Z_p)$ is transgressive. Let it correspond to $v \in H^3(L;Z_p)$. By 5.9, $P^1u = u^p$ is transgressive. Since $H^i(Y;Z_p) = 0$ for $0 < i \leq 2p$, the map

$$\delta\colon\ H^{2p}(\Omega K;Z_p) \longrightarrow H^{2p+1}(Y,\Omega K;Z_p)$$

is a monomorphism. Hence $\delta u^p = \rho^* P^1 v$ is non-zero. Hence $P^1 v$ is non-zero. This completes the proof of 5.1.

§6. The Wang Sequence.

In this section we shall prove 5.7 without using spectral sequences. We restrict ourselves to fibrations which have the covering homotopy property for all spaces (not just for triangulable spaces). (See [3] Chapter III.)

6.1 THEOREM. Let $\rho\colon E \longrightarrow X \times I$ be a fibration. Let E_t be the fibre space over X obtained by restricting E to $X \times \{t\}$ where $t \in I$. Then E_0 and E_1 are fibre homotopy equivalent fibre spaces over X.

PROOF. Let $\rho \times 1\colon E_0 \times I \longrightarrow X \times I$. Lifting this homotopy to the identity on $E_0 \times \{0\}$, we obtain a map $E_0 \times \{1\} \longrightarrow E_1$. So we have a fibre-preserving map $f\colon E_0 \longrightarrow E_1$ and similarly a fibre-preserving map $g\colon E_1 \longrightarrow E_0$. We must prove that gf is fibre homotopy equivalent to the identity and similarly for fg.

We have the map

$$\rho \times \pi \times 1\colon\ E_0 \times I \times I \longrightarrow X \times \{0\} \times I = X \times I .$$

We lift this map to E on

$$E_0 \times (\{0\} \times I \cup I \times \{0\} \cup \{1\} \times I) ,$$

by the constant lifting on $I \times \{0\}$ and using the constructions described above on $\{0\} \times I$ and $\{1\} \times I$. By the covering homotopy property we can extend the lifting to $E_0 \times I \times I$. The homotopy between gf and the identity are found by looking at the lifting restricted to $E_0 \times I \times \{1\}$. This proves the theorem.

6.2. COROLLARY. Let $f: X' \longrightarrow X$ be a map which can be contracted to a point x by a homotopy keeping $f(p) = x$ fixed. Suppose we have a fibration over X with fibre F over x. Then the induced fibration $E' \longrightarrow X'$ is fibre homotopy equivalent to the trivial fibration $X' \times F \longrightarrow X'$. The fibre homotopy equivalence maps F, the fibre over p, into F by a map which is homotopic to the identity.

PROOF. We have a map $X' \times I \longrightarrow X$ such that $X' \times 1 \cup p \times I$ is sent to x. Let E be the induced fibration over $X' \times I$. The corollary follows from 6.1.

Now suppose we have a fibration $X \longrightarrow S^n$. By 6.2 if we restrict the fibre space to any proper subspace of S^n, we have a fibration which is fibre homotopy equivalent to the trivial fibration. Let $S^n = E_+ \cup E_-$ where $E_+ \cap E_- = S^{n-1}$. Let F be the fibre over a base-point $x \in S^{n-1}$. Let X_+ be the part of the fibre space over E_+ and X_- the part over E_-. Then we have the commutative diagram

$$\begin{array}{ccccc} (E_+ \times F, S^{n-1} \times F) & \longrightarrow & (X, X_-) & \longleftarrow & (X, F) \\ \downarrow & & \downarrow & & \downarrow \\ (E_+, S^{n-1}) & \longrightarrow & (S^n, E_-) & \longleftarrow & (S^n, x) \end{array}$$

Using excision and 6.2 we easily deduce the isomorphisms

$$H^*(E_+, S^{n-1}) \otimes H^*(F) \approx H^*(E_+ \times F, S^{n-1} \times F) \approx H^*(X, X_-) \approx H^*(X, F) .$$

Hence $H^k(X,F) \approx H^{k-n}(F)$. Under this isomorphism the cohomology sequence of (X,F) becomes

$$H^k(X) \longrightarrow H^k(F) \xrightarrow{\theta} H^{k-n+1}(F) \longrightarrow H^{k+1}(X) .$$

This is the Wang sequence. We have yet to prove that θ is a derivation. We have the commutative diagram

$$\begin{array}{ccccccc} H^k(F) & \approx & H^k(X_-) & \longrightarrow & H^k(S^{n-1} \times F) & \longrightarrow & H^k(x \times F) \approx H^k(F) \\ \downarrow \delta & & \downarrow \delta & & \downarrow \delta & & \\ H^{k+1}(X,F) & \approx & H^{k+1}(X, X_-) & \approx & H^{k+1}(E_+ \times F, S^{n-1} \times F) & \approx & H^{k+1-n}(F) \end{array} .$$

The composition on the top line is the identity by the last sentence in the statement of 6.2. Hence $x \in H^k(F)$ goes to

$$u \times \theta x + 1 \times x \in H^k(S^{n-1} \times F)$$

where u generates $H^{n-1}(S^{n-1})$. Therefore if $y \in H^*(F)$, xy goes to

$$u \times (\theta x \cdot y + (-1)^{(n-1)k} x \cdot \theta y) + 1 \times xy \in H^*(S^{n-1} \times F) .$$

This shows that

$$\theta(xy) = \theta x \cdot y + (-1)^{(n-1)k} x \cdot \theta y$$

and completes the proof of 5.7.

BIBLIOGRAPHY

[1] S. P. Novikov, "Homotopy properties of Thom complexes," (dissertation to appear).

[2] C. T. C. Wall, "Generators and relations for the Steenrod algebra," Annals of Math., 72 (1960), pp. 429-444.

[3] S. Hu, "Homotopy Theory," Academic Press 1959.

[4] J. P. Serre, "Groupes d'homotopie et classes des groupes abélians," Annals of Math., 58 (1953) pp. 258-294.

[5] J. P. Serre, "Homologie singuliére des espaces fibrés," Annals of Math., 54 (1951) pp. 425-505.

[6] A. Borel and J. P. Serre, "Groupes de Lie et puissances réduites de Steenrod," Amer. J. of Math., 75 (1953), pp. 409-448.

CHAPTER VII.

Construction of the Reduced Powers

In §1 we explain how the reduced powers are a fairly natural generalization of products in cohomology groups. In §2 we define the external reduced power map P in general situation and prove some of its properties. In §3 we specialize to the case of the cyclic group of permutations of p factors, where p is a prime and the coefficient group Z_p. In §4 we use the transfer to prove some further properties of the reduced powers. In §5 we determine the reduced power of degree zero. In §6 we define P^i and Sq^i and prove all the axioms in Chapters VI and I except for the Adem relations, which will be proved in Chapter VIII.

§1. Intuitive Ideas behind the Construction.

Let K be a finite regular cell complex and let K^n be the n-fold Cartesian product. Let S(n) be the symmetric group on n elements acting as permutations of the factors of K^n. Let π be a subgroup of S(n) and let W be a π-free acyclic complex. $W \times K^n$ is a π-free complex via the diagonal action.

Apart from these definitions, an understanding of this section is not logically necessary for the understanding of what follows. In places this section is deliberately vague.

Let L be another finite regular cell complex. Let $u \in H^*(K)$ and $v \in H^*(L)$. Then we have the cross-product $u \times v \in H^*(K \times L)$. If K = L and $d: K \longrightarrow K \times K$ is the diagonal we define the cup-product

$$u \cup v = d^*(u \times v).$$

The cup-product is called an internal operation since all the cohomology classes exist in a single space K; the cross-product is called an external operation. The advantage of the cross-product is that its definition requires no choice, even on the cochain level. On the other hand, the cup-product requires a diagonal approximation $d_{\#}\colon K \longrightarrow K \otimes K$. Many difficulties experienced with the cup-product in the past arose from the great variety of choices of $d_{\#}$, any particular choice giving rise to artificial-looking formulas. Moreover, the properties of the cup-product such as the associative and commutative laws follow easily from the corresponding properties for the cross-product by applying the diagonal. The properties for the cross-products themselves are easy to prove.

Similarly we shall obtain the (internal) reduced powers as images, under an analogue of the diagonal mapping, of a certain external operation P. We shall prove many of the properties of the (internal) reduced powers by proving the corresponding properties for the external operation.

Let $W \times_\pi K^n = (W \times K^n)/\pi$ and let j be the composition (which is an embedding)

$$K^n \longrightarrow W \times K^n \longrightarrow W \times_\pi K^n .$$

The map $W \times_\pi K^n \longrightarrow W/\pi$ is a fibration with fibre K^n. Given a cohomology class u on K, we have a class $u \times \ldots \times u$ on K^n. Under suitable conditions we can extend this class in one and only one way to a class Pu in the total space $W \times_\pi K^n$ so that P is natural with respect to maps of the variable K, $P0 = 0$ and

$$j^*Pu = u \times \ldots \times u.$$

For the nth power in the sense of cup-products, we have

$$u^n = d^*(u \times \ldots \times u).$$

To define <u>reduced</u> n^{th} powers, we replace K^n by $W \times_\pi K^n$ and $u \times \ldots \times u$ by Pu. We replace $d\colon K \longrightarrow K^n$ by

$$1 \times_\pi d\colon W \times_\pi K \longrightarrow W \times_\pi K^n .$$

Now $W \times_\pi K = W/\pi \times K$. Hence

$$(1 \times_\pi d)^*Pu \in H^*(W/\pi \times K) .$$

If we are working with a field of coefficients, we can expand in $H^*(W/\pi \times K)$ by the Künneth theorem. The coefficients of the expansion of $(1 \times_\pi d)^*Pu$ which lie in $H^*(K)$ are the internal reduced n^{th} powers.

In subsequent sections we replace cohomology classes on $W \times_\pi K^n$ by equivariant cohomology classes on $W \times K^n$.

§2. Construction.

Let K be a finite regular cell complex. Suppose we are given a q-cocycle u on K with values in an abelian group G. We regard G as a complex with all components $G_r = 0$ except in dimension zero $G_0 = G$. Then we have a chain map $u\colon K \longrightarrow G$ which lowers degrees by q. Let $G^n(q)$ be the $S(n)$-complex defined as follows. It is zero in non-zero dimensions and is the n-fold tensor product G^n in dimension zero. We let $\alpha \in S(n)$ act on G^n by the product of the sign of α and the permutation of the factors of G^n if q is odd. If q is even we let α permute the factors of G^n with no sign change. Then $u^n\colon K^n \longrightarrow G^n(q)$ is an equivariant chain map which lowers degrees by nq.

Let $\varepsilon\colon W \longrightarrow Z$ be the augmentation on W. Then $\varepsilon \otimes 1 : W \otimes K^n \longrightarrow K^n$ is an equivariant chain map (using the diagonal action on $W \otimes K^n$). Therefore the composition

$$W \otimes K^n \xrightarrow{\varepsilon \otimes 1} K^n \xrightarrow{u^n} G^n(q)$$

is an equivariant chain map which lowers degrees by nq. In other words, we have an equivariant nq-cocycle on $W \otimes K^n$, which we denote by

$$Pu \in C_\pi^{nq}(W \otimes K^n; G^n(q)).$$

We now prove that if we vary u by a cohomology, then Pu varies by an equivariant cohomology.

2.1. LEMMA. There exists an equivariant map $h\colon I \otimes W \longrightarrow I^n \otimes W$ such that $h(\bar{0} \otimes w) = \bar{0}^n \otimes w$ and $h(\bar{1} \otimes w) = \bar{1}^n \otimes w$, for all $w \in W$.

PROOF. h is equivariant on $\bar{0} \otimes W$ and $\bar{1} \otimes W$. We have the equivariant acyclic carrier $W \otimes I^n$. The lemma follows from V 2.2.

2.2. LEMMA. If u and v are cohomologous q-cocycles on K with values in G, then Pu and Pv are cohomologous nq-cocycles in $C_\pi^*(W \otimes K^n; G^n(q))$— that is, they are equivariantly cohomologous.

PROOF. Now u cohomologous to v means that there is a chain homotopy of u into v, that is a chain map $D\colon I \otimes K \longrightarrow G$ lowering degrees by q, such that $D(\bar{0} \otimes \tau) = u(\tau)$ and $D(\bar{1} \otimes \tau) = v(\tau)$ for all $\tau \in K$. By 2.1 we have the following composition of equivariant chain maps

$$I \otimes W \otimes K^n \xrightarrow{h\otimes 1} I^n \otimes W \otimes K^n \xrightarrow{1\otimes\varepsilon\otimes 1} I^n \otimes K^n \xrightarrow{\text{shuf}} (I \otimes K)^n \xrightarrow{D^n} G^n(q).$$

The map shuf denotes the shuffling of the two sets of n factors with the usual sign convention. This composition gives the equivariant homotopy of Pu into Pv which shows they are equivariantly cohomologous.

The lemma shows that P induces a map (not a homomorphism in general)

$$P\colon\ H^q(K;G) \longrightarrow H^{nq}_\pi(W \otimes K^n;G^n(q)).$$

Let w be a 0-dimensional cell of W. We have a map $j\colon K^n \longrightarrow W \otimes K^n$ defined by $j(x) = w \otimes x$ for all $x \in K^n$. If L is another finite regular cell complex and $f\colon K \longrightarrow L$ is a continuous map, then by V 3.3 the equivariant continuous map $f^n\colon K^n \longrightarrow L^n$ induces a map

$$(f^n)^*\colon\ H^*_\pi(W \times L^n;G^n(q)) \longrightarrow H^*_\pi(W \times K^n;G^n(q)).$$

2.3. LEMMA. 1) j^*Pu is the n-fold cross-product $u \times \ldots \times u \in H^{nq}(K^n;G^n)$.

2) We have a commutative diagram

$$\begin{array}{ccc} H^q(L;G) & \xrightarrow{P} & H^{nq}_\pi(W \times L^n;G^n(q)) \\ \downarrow{\scriptstyle f^*} & & \downarrow{\scriptstyle (f^n)^*} \\ H^q(K;G) & \xrightarrow{P} & H^{nq}_\pi(W \times K^n;G^n(q)). \end{array}$$

PROOF. 1) follows immediately by the definitions on the cochain level of P and of cross-products.

We reduce the proof of 2) to the case where f is proper by using V 3.1 and V 3.3. Let C be the minimal carrier of f. Then the carrier from K^n to L^n which sends $\sigma_1 \times \ldots \times \sigma_n$ to $C(\sigma_1) \times \ldots \times C(\sigma_n)$ is an acyclic equivariant carrier for f^n. Therefore, if $f_\#\colon K \longrightarrow L$ is a chain approximation to f, we can use $1 \otimes (f_\#)^n$ as our equivariant map $W \otimes K^n \longrightarrow W \otimes L^n$. We have a commutative diagram

$$\begin{array}{ccccc} W \otimes K^n & \xrightarrow{\varepsilon \otimes 1} & K^n & \xrightarrow{(f^\# u)^n} & G^n(q) \\ \downarrow{\scriptstyle 1 \otimes (f_\#)^n} & & \downarrow{\scriptstyle (f_\#)^n} & & \| \\ W \otimes L^n & \xrightarrow{\varepsilon \otimes 1} & L^n & \xrightarrow{u^n} & G^n(q). \end{array}$$

The lemma follows.

2.4. REMARK. If $n = p$ and $G = Z_p$ and π is the subgroup of $S(p)$ which permutes the factors of K^p cyclically, then P is characterized by the properties in 2.3 and the fact that $P0 = 0$. (This can be proved by the methods of VIII §3.)

Let $\pi \subset \rho \subset S(n)$ and let V and W be respectively a ρ-free and a π-free acyclic complex.

2.5. LEMMA. The following diagram is commutative

$$\begin{array}{ccc} & \overset{P}{\nearrow} & H^{nq}_{\pi}(W \times K^n; G^n(q)) \\ H^q(K;G) & & \uparrow \\ & \underset{P}{\searrow} & H^{nq}_{\rho}(V \times K^n; G^n(q)) \end{array}$$

where the map on the right is induced as in V 3.3. It follows that P is independent of the choice of W.

PROOF. Let $g_{\#}\colon W \longrightarrow V$ be an equivariant chain map. The diagram

$$\begin{array}{cccc} W \otimes K^n & \overset{\varepsilon\otimes 1}{\searrow} & & \\ g_{\#}\otimes 1 \downarrow & & K^n \xrightarrow{u^n} G^n(q) \\ V \otimes K^n & \underset{\varepsilon\otimes 1}{\nearrow} & & \end{array}$$

is commutative. The lemma follows.

Let $u \in H^q(K;G)$ and $v \in H^r(L;F)$ where K and L are finite regular cell complexes and G and F are abelian groups. We have $Pu \in H^{nq}_{\pi}(W \times K^n; G^n(q))$ and $Pv \in H^{nr}_{\pi}(W \times L^n; F^n(r))$. By V 4.2, we have a cross-product

$$Pu \times Pv \in H^{nq+nr}_{\pi \times \pi}(W \times W \times K^n \times L^n;\ G^n(q) \otimes F^n(r))$$

where $\pi \times \pi$ acts on $W \times W \times K^n \times L^n$ by the formula

$$(\alpha,\beta)(v_1,v_2,x,y) = (\alpha v_1, \beta v_2, \alpha x, \beta y)$$

for all $\alpha,\beta \in \pi$, $v_1,v_2 \in W$, $x \in K^n$ and $y \in L^n$. We also have

$$u \times v \in H^{q+r}(K \times L; G \otimes F) \quad \text{and}$$
$$P(u \times v) \in H^{n(q+r)}_{\pi}(V \times (K \times L)^n; (G \otimes F)^n(q + r))$$

where V is a π-free acyclic complex.

We have a map of geometric triples

$$\lambda\colon\ (\pi, (G \otimes F)^n(q + r), (K \times L)^n) \longrightarrow (\pi \times \pi, G^n(q) \otimes F^n(r), K^n \times L^n)$$

defined as follows: $\lambda_1: \pi \longrightarrow \pi \times \pi$ is such that $\lambda_1(\alpha) = (\alpha,\alpha)$ for all $\alpha \in \pi$;

$$\lambda_2: G^n(q) \otimes F^n(r) \longrightarrow (G \otimes F)^n(q + r)$$

is the obvious isomorphism which shuffles the two sets of n variables; the map $(K \times L)^n \longrightarrow K^n \times L^n$ unshuffles the two sets of n variables. By V 3.3 we have a map

$$\lambda^*: H^*_{\pi\times\pi}(W \times W \times K^n \times L^n; G^n(q) \otimes F^n(r)) \longrightarrow H^*_\pi(V \times (K \times L)^n; (G \otimes F)^n(q + r)).$$

2.6. LEMMA. $\lambda^*(Pu \times Pv) = (-1)^{n(n-1)qr/2}P(u \times v)$.

PROOF. According to 2.5 we may take V to be an arbitrary π-free acyclic complex. Let $V = W \times W$ with the diagonal action. We have the commutative diagram of equivariant chain maps

$$\begin{array}{ccc}
(W \otimes W) \otimes (K \otimes L)^n & \xrightarrow{1\otimes\lambda_\#} & W \otimes W \otimes K^n \otimes L^n \\
\downarrow{\scriptstyle \varepsilon\otimes 1} & & \downarrow{\scriptstyle \varepsilon\otimes\varepsilon\otimes 1\otimes 1} \\
(K \otimes L)^n & \xrightarrow{\lambda_\#} & K^n \otimes L^n \\
\downarrow{\scriptstyle (u\otimes v)^n} & & \downarrow{\scriptstyle u^n\otimes v^n} \\
(G \otimes F)^n(q + r) & \xrightarrow{\mu} & G^n(q) \otimes F^n(r)
\end{array}$$

where μ is $(-1)^{n(n-1)rq/2}$ times the inverse of λ_2. The left side of the diagram gives $P(u \times v)$ and the right side gives $Pu \times Pv$. This proves the lemma.

§3. Cyclic Reduced Powers.

Now let $n = p$, a prime, and let $G = Z_p$. Then $G^p(q)$ is isomorphic to Z_p as an abelian group. $S(p)$ acts on $Z_p = G^p(q)$ by the sign of the permutation if q is odd, and trivially if q is even. In the notation of V §6, $G^p(q) = Z_p^{(q)}$.

Let $\pi \subset S(p)$ be the cyclic group of order p, generated by the permutation T which sends i to $(i + 1) \bmod p$. The sign of this permutation is $(-1)^{p-1}$. Since $(-1)^{p-1} \equiv 1 \pmod p$, $Z_p^{(q)}$ is a trivial π-module.

3.1. LEMMA. Let K be a finite regular cell complex with no π-action. Then

$$H_\pi^*(W \times K; Z_p) \approx H^*(W/\pi \times K; Z_p)$$

and this isomorphism is natural for maps of K.

Let $d\colon K \longrightarrow K^p$ be the diagonal map. Then d is equivariant, if $S(p)$ acts on K^p by permuting its factors. By V 3.3 we have an induced map

$$d^*\colon H_\pi^*(W \times K^p, Z_p^{(q)}) \longrightarrow H_\pi^*(W \times K; Z_p^{(q)}) .$$

Since $Z_p^{(q)}$ is a trivial π-module, we can replace $Z_p^{(q)}$ by Z_p. So, if $u \in H^q(K;Z_p)$, we have by 3.1 and the Künneth formula

3.2. DEFINITION. $d^*Pu = \Sigma_k w_k \times D_k u$

where $w_k \in H^k(W/\pi;Z_p)$ are the elements of V 5.2, and this defines

$$D_k\colon H^q(K;Z_p) \longrightarrow H^{pq-k}(K;Z_p).$$

(Note that we have not yet shown that D_k is a homomorphism.)

Let $f\colon K \longrightarrow L$ be a continuous map between two finite regular cell complexes with no group action.

3.3. LEMMA. For each k, $f^*D_k = D_k f^*$.

PROOF. We have $df = f^p d$. Hence the following diagram is commutative (by V 3.3)

$$\begin{array}{ccc} H_\pi^{pq}(W \times L^p;Z_p) & \xrightarrow{d^*} & H_\pi^{pq}(W \times L;Z_p) \\ \downarrow (f^p)^* & & \downarrow f^* \\ H_\pi^{pq}(W \times K^p;Z_p) & \xrightarrow{d^*} & H_\pi^{pq}(W \times K;Z_p) \end{array}$$

Applying the commutative diagram of 2.3 to the left and the isomorphsims of 3.1 to the right of this diagram, the lemma follows.

3.4. LEMMA. $D_0 u = u^p$.

PROOF. Let w be a 0-cell of W. Let $d_\#\colon K \longrightarrow K^p$ be a diagonal approximation. We have a commutative diagram

$$\begin{array}{ccc} K & \xrightarrow{j} & W \otimes K \\ d_\# \downarrow & & \downarrow 1\otimes d_\# \\ K^p & \xrightarrow{j} & W \otimes K^p \end{array}$$

where $jx = w \otimes x$. Now

$$\begin{aligned} D_0 u &= j^*(\Sigma_k w_k \times D_k u) \\ &= j^* d^* Pu \\ &= d^* j^* Pu \\ &= d^*(u \times \ldots \times u) \quad \text{by 2.3} \\ &= u^p . \end{aligned}$$

3.5. LEMMA. Let $u \in H^q(K;Z_p)$ and let $p > 2$. If q is even $D_j u = 0$ unless $j = 2m(p-1)$ or $2m(p-1)-1$ for some non-negative integer m. If q is odd, $D_j u = 0$ unless $j = (2m+1)(p-1)$ or $(2m+1)(p-1)-1$ for some non-negative integer m.

PROOF. With notation as in V §6, let γ^* be the automorphism of $H^*(W \times L;Z_p^{(q)})$ induced by $\gamma \in \rho$ as in V 3.4, where L is a finite regular cell complex on which ρ acts. Let V be a ρ-free acyclic complex. By 2.5, V 3.3 and V 3.4 we have the commutative diagram

$$\begin{array}{ccccccc} & \nearrow^{P} & H_\rho^{pq}(V \times K^p;Z_p^{(q)}) & \xrightarrow{d^*} & H_\rho^{pq}(V \times K;Z_p^{(q)}) & \xrightarrow{\gamma^*=1} & H_\rho^{pq}(V \times K;Z_p^{(q)}) \\ H^q(K;Z_p) & & \downarrow & & \downarrow & & \downarrow \\ & \searrow_{P} & H_\pi^{pq}(W \times K^p;Z_p^{(q)}) & \xrightarrow{d^*} & H_\pi^{pq}(W \times K;Z_p^{(q)}) & \xrightarrow{\gamma^*} & H_\pi^{pq}(W \times K;Z_p^{(q)}) . \end{array}$$

The lemma follows from V 6.1 and V 6.2.

§4. The Transfer.

We have defined the transfer in V §7. Let $d^*: H^*(W \times K^p;Z_p) \longrightarrow H^*(W \times K;Z_p)$ the map induced by the diagonal $d: K \longrightarrow K^p$.

4.1 LEMMA. Let $\tau: H^*(W \otimes K^p;Z_p) \longrightarrow H_\pi^*(W \otimes K^p;Z_p)$ denote the transfer. Then $d^*\tau = 0$.

PROOF. We have a commutative diagram

$$\begin{array}{ccccc} & & H^*(W \otimes K^p;Z_p) & \xrightarrow{\tau} & H_\pi^*(W \otimes K^p;Z_p) \\ & & \downarrow d^* & & \downarrow d^* \\ H_\pi^*(W \otimes K;Z_p) & \xrightarrow{i^*} & H^*(W \otimes K;Z_p) & \xrightarrow{\tau} & H_\pi^*(W \otimes K;Z_p) . \end{array}$$

Since W is acyclic and $H^0_\pi(W;Z_p) \longrightarrow H^0(W;Z_p)$ is onto,

$$i^*: \ H^n_\pi(W \otimes K;Z_p) \longrightarrow H^n(W \otimes K;Z_p)$$

is also onto. By V 7.1 $\tau i^* = 0$. The lemma follows.

4.2. LEMMA. If π is the group of cyclic permutations and $P: \ H^q(K;Z_p) \longrightarrow H^{pq}_\pi(W \times K^p;Z_p)$ then d^*P is a homomorphism.

PROOF. Let u and v be q-cocycles on K. Then $P(v+u) - Pu - Pv$ is given by the chain map

$$W \otimes K^p \xrightarrow{\varepsilon \otimes 1} K^p \xrightarrow{(u+v)^p - u^p - v^p} Z_p \ .$$

According to 4.1, we need only show that this cocycle is in the image of the transfer. It will be sufficient to show that $(u+v)^p - u^p - v^p$ is in the image of a cocycle under

$$\tau: \ C^*(K^p;Z_p) \longrightarrow C^*_\pi(K^p;Z_p)$$

since $\varepsilon \otimes 1$ is an equivariant map.

Now $(u+v)^p - u^p - v^p$ is the sum of all monomials which contain k factors u and $(p-k)$ factors v, where $1 \leq k \leq p-1$. Now π permutes such factors freely. Let us choose a basis consisting of monomials whose permutations under π give each monomial exactly once. Let z be the sum of the monomials in the basis. Then $\tau z = (u+v)^p - u^p - v^p$. Also z is a cocycle in K^p since each monomial is a cocycle. The lemma follows.

4.3. COROLLARY. For each k,

$$D_k: \ H^q(K;Z_p) \longrightarrow H^{pq-k}(K;Z_p)$$

is a homomorphism.

4.4. LEMMA. If $u \in H^q(K;Z_p)$ then $D_k u = 0$ for $k > (p-1)q$ and $D_{(p-1)q} u = a_q u$ where $a_q \in Z_p$ is a constant which is independent of u and K.

PROOF. Let K^q be the q-skeleton of K. Then

$$i^*: \ H^r(K) \longrightarrow H^r(K^q)$$

is a monomorphism for $r \leq q$. By 3.3 we can therefore assume that K is q-dimensional. Let $u_0 \in H^q(S^q;Z_p)$ be the class dual to S^q. There is a map $f\colon K \longrightarrow S^q$ such that $f^*u_0 = u$: we let $f(K^{q-1})$ be a point and map each q-cell of K into S^q with degree given by v, a cocycle representative for u. By 3.3 we can assume that $K = S^q$ and $u = u_0$. The second part of the lemma follows. If $k > (p-1)q$, then the only possibility remaining for $D_k u$ to be non-zero and $k > (p-1)q$ is that $k = pq$ and $q > 0$. Let $j\colon s \longrightarrow S^q$ be the inclusion of a point s in S^q. Then j^* is an isomorphism in dimension zero and $j^*u = 0$.

$$\begin{aligned} j^*D_{pq}u &= D_{pq}j^*u \qquad \text{by } 3.3 \\ &= D_{pq}0 \\ &= 0 \qquad\qquad \text{by } 4.3. \end{aligned}$$

This proves the lemma.

4.5. LEMMA. Let β be the Bockstein operator associated with the exact sequence

$$0 \longrightarrow Z_p \longrightarrow Z_{p^2} \longrightarrow Z_p \longrightarrow 0.$$

Then $\beta d^*Pu = 0$ for $p > 2$ or q even.

PROOF. Since $\beta d^* = d^*\beta$, 4.1 shows that we have only to prove that βPu is in the image of the transfer. Let v be an integral cochain on K represented $u \in H^q(K;Z_p)$. Then $\delta v = pz$ where z is an integral $(q+1)$-cocycle, and z represents $\beta u \in H^{q+1}(K;Z_p)$. The cochain v^p is an integral cochain on K^p whose cohomology class we denote by $\{v^p\} \in H^{pq}_\pi(K^p;Z_p)$. Let $\varepsilon \otimes 1\colon W \otimes K^p \longrightarrow K^p$. Then

$$\beta Pu = \beta(\varepsilon \otimes 1)^*\{v^p\} = (\varepsilon \otimes 1)^*\beta\{v^p\}.$$

Since τ commutes with $(\varepsilon \otimes 1)^*$, it will be sufficient to show that $\beta\{v^p\}$ is in the image of τ.

$$\begin{aligned} \delta v^p &= \Sigma_{s=0}^{p-1} (-1)^{qs}\, v^s(\delta v)v^{p-s-1} \\ &= p\, \Sigma_{s=0}^{p-1} (-1)^{qs}\, v^s\, z\, v^{p-s-1} \\ &= p\, \Sigma_{\alpha\in\pi} (-1)^{qs(p-1)}\, \alpha(zv^{p-1}) \\ &= p\, \tau(zv^{p-1}) \end{aligned}$$

since either $p-1$ or q is even. Since v is a mod p cocycle, zv^{p-1} is a mod p cocycle. The above argument shows that $\tau(zv^{p-1})$ represents

$\beta\{v^p\}$, and the proof of the lemma is complete.

4.6. COROLLARY. If $p > 2$ or $q = \dim u$ is even then $\beta D_0 u = 0$, $\beta D_{2k} u = D_{2k-1} u, \beta D_{2k-1} u = 0$.

PROOF. By 3.2 and 4.5

$$\beta(\Sigma_k w_k \times D_k u) = 0$$

From V 5.2, $\beta w_{2j} = 0$ and $\beta w_{2j+1} = -w_{2j+2}$ $(j \geq 0)$. Hence

$$\Sigma_{k \geq 0} w_{2k} \times \beta D_{2k} u - \Sigma_{k \geq 0} w_{2k+1} \times \beta D_{2k+1} u - \Sigma_{k \geq 1} w_{2k} \times D_{2k-1} u = 0$$

The lemma follows by comparing coefficients of w_k.

4.7. LEMMA. Let $u \in H^r(K;Z_p)$ and $v \in H^s(L;Z_p)$. If $p > 2$ then

$$D_{2k}(u \times v) = (-1)^{p(p-1)rs/2} \Sigma_{j=0}^{k} D_{2j} u \times D_{2k-2j} v$$

If $p = 2$, $D_k(u \times v) = \Sigma_{j=0}^{k} D_j u \times D_{k-j} v$.

PROOF. The map of geometric triples λ, used in 2.6, takes the form

$$\lambda: (\pi, Z_p, (K \times L)^p) \longrightarrow (\pi \times \pi, Z_p, K^p \times L^p).$$

We have a commutative diagram of maps of geometric triples

$$\begin{array}{ccc}
(\pi, Z_p, (K \times L)^p) & \xrightarrow{\lambda} & (\pi \times \pi, Z_p, K^p \times L^p) \\
\uparrow d & & \uparrow d' \\
(\pi, Z_p, K \times L) & \xrightarrow{d_1} & (\pi \times \pi, Z_p, K \times L)
\end{array}$$

where d is induced by the diagonal on $K \times L$, d_1 by the diagonal on π and d' by combining the diagonals on K and L.

Let W be a π-free acyclic complex. Then $W \times W$ is a $(\pi \times \pi)$-free acyclic complex. From the above diagram and V 3.3 we have a commutative diagram

$$\begin{array}{ccc}
H_\pi^*(W \times (K \times L)^p; Z_p) & \longleftarrow & H_{\pi\times\pi}^*((W \times W) \times (K^p \times L^p); Z_p) \\
\downarrow d^* & & \downarrow (d')^* \\
H_\pi^*(W \times K \times L; Z_p) & \longleftarrow & H_{\pi\times\pi}^*(W \times W \times K \times L; Z_p)
\end{array}$$

According to V 4.2, $Pu \times Pv$ is an element in the group on the upper right of the diagram. We have

$$H^*_{\pi\times\pi}(W \times W \times K \times L;Z_p) \approx H^*(W/\pi \times W/\pi \times K \times L;Z_p) .$$

It is easy to see that under this isomorphsim we have by 3.2

$$(d')^*(Pu \times Pv) = \Sigma_{j,\ell} (-1)^{\ell(pr-j)} w_j \times w_\ell \times D_ju \times D_\ell v.$$

Applying $(d_1)^*$ to each side of this equation, and using the commutative diagram, we obtain

$$d^*\lambda^*(Pu \times Pv) = \Sigma_{j,\ell} (-1)^{\ell(pr-j)} w_jw_\ell \times D_ju \times D_\ell v.$$

Also $d^*P(u\times v) = \Sigma_k w_k \times D_k(u\times v)$. The lemma follows from 2.6 and V 5.2.

§5. Determination of $D_{q(p-1)}$.

We know from 4.4 that for each q there is a constant $a_q \in Z_p$, such that

$$D_{q(p-1)}u = a_qu .$$

5.1. LEMMA. $a_q = (-1)^r a_1^q$ where $r = p(p-1)q(q-1)/4$.

PROOF. The lemma is proved by induction on q. It is true for $q = 0$ by 4.4.

Let $u \in H^{q-1}(K;Z_p)$ be non-zero and let v be a generator of $H^1(S^1;Z_p)$. Then $u \times v \in H^q(K \times S^1;Z_p)$ is non-zero. By 4.4 $D_jv = 0$ unless $j = p-1$. By 4.7

$$\begin{aligned} D_{q(p-1)}(u \times v) &= (-1)^{p(p-1)(q-1)/2} D_{(q-1)(p-1)}u \times D_{p-1}v \\ &= (-1)^{p(p-1)(q-1)/2} a_{q-1}a_1(u \times v) . \end{aligned}$$

Hence $a_q = (-1)^{p(p-1)(q-1)/2} a_{q-1}a_1$. The lemma follows by induction.

In order to complete the determination of $D_{q(p-1)}$, we must find a_1. This is done by appealing directly to the definition in the case of S^1.

Suppose K is a finite regular cell complex and $u \in H^q(K;Z_p)$. Then $\Sigma_j w_j \times D_ju$ is represented by the composition

$$W \times K \xrightarrow{\varepsilon\otimes d_\#} W \times K^p \xrightarrow{\varepsilon\otimes 1} K^p \xrightarrow{u^p} Z_p ,$$

where $d_\#$ is a diagonal approximation. By V 2.2 any two equivariant chain maps $W \otimes K \longrightarrow K^p$, carried by the diagonal carrier, are equivariantly homotopic.

Hence, in order to find D_{p-1} on a 1-dimensional class, we need only find an equivariant chain map

$$\varphi: \; W \otimes S^1 \longrightarrow (S^1)^p$$

carried by the diagonal carrier. We make S^1 into a regular complex by breaking it into two intervals J_1 and J_2 such that $\partial J_1 = A - B$ and $\partial J_2 = A - B$. Then the fundamental homology class of S^1 is $J_1 - J_2$.

Let W be the complex of V §5. We define

$$\phi(e_0 \otimes A) = A^p \;\; ; \; \phi(e_0 \otimes B) = B^p \;\; ;$$

$$\varphi(e_j \otimes A) = \varphi(e_j \otimes B) = 0 \quad \text{for} \quad j > 0.$$

In fact φ is uniquely determined thus far by the carrier. We need only extend the definition of φ to an equivariant chain map

$$\varphi: \; W \otimes I \longrightarrow I^p$$

where $\partial I = B - A$, and this will give a formula $W \otimes S^1$ by taking first $J_1 = I$ and then $J_2 = I$.

We define

$$\varphi(e_{2i} \otimes I) = i! \, \Sigma(A^{\alpha_0} IB^{\beta_0})(IA^{\alpha_1} IB^{\beta_1}) \ldots (IA^{\alpha_i} IB^{\beta_i})$$

where the summation extends over all sequences (α, β) such that $\Sigma_{j=0}^{i} (\alpha_j + \beta_j) = p - 2i - 1$; and

$$\varphi(e_{2i+1} \otimes I) = i! \, \Sigma(IA^{\alpha_0} IB^{\beta_0}) \ldots (IA^{\alpha_i} IB^{\beta_i})$$

where the summation extends over all sequences (α, β) such that $\Sigma_{j=0}^{i} (\alpha_j + \beta_j) = p - 2i - 2$.

The problem now is to show that φ is a chain map. We do this by using a contracting homotopy in I^p. Let s be the contracting homotopy in I given by $sA = 0$, $sB = I$, $sI = 0$. Then if $\varepsilon: I \longrightarrow A$ is the augmentation

$$s\partial + \partial s = 1 - \varepsilon .$$

We define a contracting homotopy S in I^p by the usual formula

$$S = s \otimes 1^{p-1} + \Sigma_{r=1}^{p-1} \varepsilon^r \otimes s \otimes 1^{p-r-1} + \varepsilon^{p-1} \otimes s$$

Then

$$\partial S + S\partial = 1^p - \varepsilon^p .$$

The following formulas will help us to evaluate S. Let C be any chain in I^r for some $r \geq 0$. Then we easily see that

(i) $S(A^p) = 0$ (ii) $S(B^p) = \sum_{r=0}^{p-1} A^r I B^{p-r-1}$

(iii) $S(A^k IC) = 0 \quad (k \geq 0)$ (iv) $S(B^t A^s IC) = \sum_{r=0}^{t-1} A^r I B^{t-r-1} A^s IC$ $(t \geq 1,\ s \geq 0)$.

We shall prove the following formulas

a) $\varphi(e_{2i+1} \otimes I) = S\varphi\, \partial(e_{2i+1} \otimes I)$;

b) $\varphi(e_{2i} \otimes I) = S\varphi\, \partial(e_{2i} \otimes I)$;

c) $\varphi(e_i \otimes A) = 0 = S\varphi\, \partial(e_i \otimes A)$ if $i > 0$,

$\varphi(e_i \otimes B) = 0 = S\varphi\, \partial(e_i \otimes B)$ if $i > 0$.

Let $\Delta = T - 1$, where T is the element of π which sends i to $i + 1 \pmod p$. Then

$$\begin{aligned} S\varphi\, \partial(e_{2i+1} \otimes I) &= S\varphi(\Delta(e_{2i} \otimes I)) \\ &= S\Delta\varphi(e_{2i} \otimes I) \\ &= i!\ S\ \Sigma\, (A^{\alpha_0} IB^{\beta_0})(IA^{\alpha_1} IB^{\beta_1})\dots(IA^{\alpha_i} IB^{\beta_i}) . \end{aligned}$$

By (iii), terms with $\beta_i = 0$ make no contribution. If $\beta_i > 0$, let $\beta_i' = \beta_i - 1$. Then by (iii) the above expression is equal to

$$i!\ S\ \Sigma_{\beta_i > 0}\, (BA^{\alpha_0} IB^{\beta_0})(IA^{\alpha_i} IB^{\beta_1})\dots(IA^{\alpha_i} IB^{\beta_i'}) .$$

By (iv) this expression is equal to

$$i!\ \Sigma_{\beta_i > 0}\, (IA^{\alpha_0} IB^{\beta_0})(IA^{\alpha_1} IB^{\beta_1})\dots(IA^{\alpha_i} IB^{\beta_i'}) .$$

This summation extends over all sequences (α, β) such that $\Sigma\,(\alpha_j + \beta_j) = p - 2i - 1$ and $\beta_i > 0$. Therefore the expression is equal to $\varphi(e_{2i+1} \otimes I)$. This proves a).

To prove b) we note that if $i = 0$ then

$$\begin{aligned} S\varphi\, \partial(e_0 \otimes I) &= S\varphi(e_0 \otimes B - e_0 \otimes A) \\ &= S(B^p - A^p) \\ &= \Sigma\, A^r IB^{p-r-1} \\ &= \varphi(e_0 \otimes I) . \end{aligned}$$

Let $N = 1 + T + \dots + T^{p-1}$. If $i > 0$ then

$$\begin{aligned} S\varphi\, \partial(e_{2i} \otimes I) &= S\varphi N(e_{2i-1} \otimes I) = SN\varphi(e_{2i-1} \otimes I) \\ &= (i-1)!\ SN\ \Sigma\, (IA^{\alpha_0} IB^{\beta_0})\dots(IA^{\alpha_{i-1}} IB^{\beta_{i-1}}) . \end{aligned}$$

By (iii) the only terms which make a contribution are those which begin with B. The expression is therefore equal to

$$(i-1)!\ S\ \Sigma_{(\alpha,\beta)}\ \Sigma_{j=0}^{i-1}\ \Sigma_{r=1}^{\beta_j}\ (B^r IA^{\alpha_{j+1}} IB^{\beta_{j+1}})(IA^{\alpha_{j+2}} IB^{\beta_{j+2}})\ldots$$
$$\ldots (IA^{\alpha_{j-1}} IB^{\beta_{j-1}})(IA^{\alpha_j} IB^{\beta_j - r})$$

where the subscripts k in α_k and β_k are taken mod i. By (iv) this is equal to

$$(i-1)!\ \Sigma_{(\alpha,\beta)}\ \Sigma_{j=0}^{i-1}\ \Sigma_{r=1}^{\beta_j}\ \Sigma_{t=1}^{r-1}\ (A^t IB^{r-t-1})(IA^{\alpha_{j+1}} IB^{\beta_{j+1}})\ldots(IA^{\alpha_j} IB^{\beta_j - r})$$

$$= (i-1)!\ \Sigma_{j=0}^{i-1}\ \varphi(e_{2i} \otimes I)/i!$$

$$= \varphi(e_{2i} \otimes I).$$

This proves b). Formula c) follows from the definition of φ.

From a), b) and c) we see that if c is a chain in $W \otimes I$ and $\dim c \geq 1$, then $\varphi c = S\varphi\,\partial c$.

5.2. LEMMA. φ is a chain map.

PROOF. We prove this by induction on the dimension. It is immediate in dimension 0. In dimension 1 we have

$$\begin{aligned} \varphi\partial(e_1 \otimes A) &= \varphi\Delta(e_0 \otimes A) \\ &= \Delta\varphi(e_0 \otimes A) \\ &= \Delta A^p \\ &= 0\ . \end{aligned}$$

Also $\partial\varphi(e_1 \otimes A) = 0$. Similarly $\varphi\partial(e_1 \otimes B) = 0 = \partial\varphi(e_1 \otimes B)$.
$\partial\varphi(e_0 \otimes I) = \partial\ \Sigma_{r=0}^{p-1}\ A^r IB^{p-r-1} = B^p - A^p = \varphi\partial(e_0 \otimes I)$.
This proves the lemma in dimension 1.

If $\dim c \geq 2$, then

$$\partial\varphi c = \partial S\varphi\partial c = (1 - S\partial)\ \varphi\partial c = \varphi\partial c$$

since the induction hypothesis tells us that $\partial\varphi\partial c = \varphi\partial\partial c = 0$. This proves the lemma.

Let $m = (p-1)/2$ if $p > 2$.

5.3. LEMMA. $a_1 = (-1)^m\, m!$ if $p > 2$. $a_1 = 1$ if $p = 2$.

PROOF. Let u be the cocycle of S^1 which has value 1 on J_1 and 0 on J_2. Then u generates $H^1(S^1;Z_p)$. We have

$$\begin{aligned}(w_{p-1} \otimes D_{p-1}u)(e_{p-1} \otimes (J_1 - J_2)) &= (w_{p-1} \cdot e_{p-1})[D_{p-1}u \cdot (J_1 - J_2)] \\ &= a_1,\end{aligned}$$

and

$$(w_{p-1} \otimes D_{p-1}u)(e_{p-1} \otimes (J_1 - J_2)) = u^p \cdot \varphi(e_{p-1} \otimes (J_1 - J_2)).$$

If $p = 2$ then

$$\varphi(e_1 \otimes (J_1 - J_2)) = J_1^2 - J_2^2.$$

Therefore $a_1 = 1$.

If $p > 2$, then $(p-1)$ is even and

$$\varphi(e_{p-1} \otimes (J_1 - J_2)) = m!\,(J_1^p - J_2^p).$$

Therefore $a_1 = m!\, u^p \cdot J_1^p = (-1)^{p(p-1)/2}\, m!$ This proves the lemma.

Combining 5.1 and 5.3 we obtain

5.4 THEOREM. Let $q > 0$ and let $u \in H^q K;Z_p)$. Then $D_{q(p-1)}u = a_q u$ where $a_q = 1$ if $p = 2$ and

$$a_q = (-1)^{mq(q+1)/2}(m!)^q \quad \text{if } p > 2.$$

§6. The Reduced Powers P^i and Sq^i.

6.1. DEFINITION. Let K be a finite regular cell complex and let $u \in H^q(K;Z_p)$. If $p > 2$, let $m = (p-1)/2$. We define

$$P^i u = (-1)^r (m!)^q D_{(q-2i)(p-1)}u$$

where $r = i + m(q^2 + q)/2$. If $p = 2$, we define

$$Sq^i u = D_{q-i}u .$$

Restricting ourselves for the moment (in VIII §2 the restrictions are removed) to the absolute cohomology of finite regular cell complexes we have

6.2. THEOREM. The P^i satisfy all the axioms in VI §1 (except for the Adem relations which will be proved in Chapter VIII).

We divide the proof into a number of lemmas.

6.3. LEMMA. $(m!)^2 \equiv (-1)^{m+1} \bmod p.$

PROOF. By Wilson's Theorem, $(p-1)! \equiv -1$. Therefore

$$\begin{aligned} -1 \equiv (p-1)! &= 1.2 \ldots \left(\tfrac{p-1}{2}\right) . \left(\tfrac{p+1}{2}\right) \ldots (p-1) \\ &\equiv 1.2 \ldots \left(\tfrac{p-1}{2}\right)\left[- \tfrac{(p-1)}{2}\right] \ldots (-2)(-1) \\ &\equiv (m!)^2 (-1)^m . \end{aligned}$$

The lemma follows.

6.4. LEMMA. $P^0 = 1$.

PROOF. Let $\dim u = q$. Then by 6.1

$$P^0 u = (-1)^r (m!)^{-q} D_{q(p-1)} u$$

where $n = m(q^2 + q)/2$. By 5.4, $P^0 u = u$.

6.5. LEMMA. Cartan Formula. If $u \in H^r(K)$ and $v \in H^q(L)$ then

$$P^k(u \times v) = \Sigma_{s+t=k} P^s u \times P^t v .$$

PROOF.

$$P^s u \times P^t v = (-1)^n (m!)^{-r-q} D_{(q-2s)(p-1)} u \times D_{(r-2t)(p-1)} v$$

where $n = s + t + m[q^2 + q + r^2 + r]/2$. Therefore

$$\begin{aligned} \Sigma_{s+t=k} P^s u \times P^t v &= (-1)^n (m!)^{-r-q} \Sigma_{s+t=k} D_{(q-2s)(p-1)} u \times D_{(r-2t)(p-1)} v \\ &= (-1)^{mrq+n} (m!)^{-r-q} D_{(r+q-2k)(p-1)} (u \times v) \end{aligned}$$

by 4.7, 4.4 and 3.5.

Now $mrq + n = k + m[(r + q)^2 + (r + q)]/2$. The lemma follows by 6.1.

6.6. LEMMA. If $\dim u = 2k$, then $P^k u = u^p$.

PROOF. $P^k u = (-1)^r (m!)^{-2k} D_0 u$

where $r = k + m(4k^2 + 2k)/2 \equiv k(m + 1) \bmod 2$. By 6.3

$$(m!)^{-2k} \equiv (-1)^{k(m+1)} \bmod p.$$

The lemma follows from 3.4.

Combining the lemmas we obtain 6.2.

Restricting ourselves for the moment (in VIII §2 the restrictions are removed) to absolute cohomology of finite regular cell complexes we have

6.7. THEOREM. The Sq^i satisfy the axioms of I §1 (except for the

Adem relations which we prove in Chapter VIII).

PROOF. The proof of Axioms 1)-5) is very similar to the proof of 6.2, except that we do not have to worry about coefficients in Z_p. We have only to prove that $\beta = Sq^1$. Now if $\dim u = 2q$ then by 4.6

$$Sq^1 u = D_{2q-1} u = \beta D_{2q} u = \beta Sq^0 u = \beta u.$$

In order to complete the proof of the theorem we prove the following lemma.

6.8. LEMMA. If $p = 2$ let R be a sum of compositions of the form β or Sq^i $(i = 0,1,2,\ldots)$. If p is an odd prime, let R be a sum of compositions of cohomology operations of the form β or P^i. Let n_j be a sequence of integers strictly increasing with j, and let $Ru = 0$ for any cohomology class of dimension n_j. Then Ru is zero for all cohomology classes.

PROOF. Let R be zero on classes of dimension r. We shall prove that $Ru = 0$ for all classes of dimension $(r-1)$. Let $v \in H^1(S^1;Z_p)$ be the generator. Then the only cohomology operation, amongst those in the statement of the lemma, which is non-zero is the identity (P^0 or Sq^0). By the Cartan formula

$$R(u \times v) = Ru \times v .$$

Since $\dim (u \times v) = r$, we have $Ru \times v = 0$ and hence $Ru = 0$. This proves the lemma and also completes the proof of 6.7.

BIBLIOGRAPHY

[1]. N. E. Steenrod, Products of cocycles and extensions of mappings, Ann. of Math., 48 (1947), pp. 290-320.

[2]. ______________, Homology groups of symmetric groups and reduced power operations, Proc. Nat. Acad. Sci. USA., 39 (1953), pp. 213-223 .

[3]. ______________, Cohomology operations derived from the symmetric group, Comment. Math. Helv., 31 (1957), pp. 195-218.

[4]. Emery Thomas, The generalized Pontrjagin cohomology operations and rings with divided powers, Memoirs Amer. Math. Soc., 27 (1957).

CHAPTER VIII.

The Relations of Adem and The Uniqueness Theorem.

In §1 we shall prove that the operations P^i and Sq^i defined in Chapter VII satisfy the Adem relations. In §2 we shall show how to extend the domain of definition of the reduced powers so that they operate in relative cohomology and in the Čech and singular theories. In §3 we prove that the reduced powers are uniquely determined by the first five axioms.

§1. The Adem Relations.

Let $S(p^2)$ be the symmetric group on p^2 elements, namely the ordered pairs (i,j) with $i,j \in Z_p$, arranged in a matrix with (i,j) in the ith row and j^{th} column. Let $\alpha(i,j) = (i+1,j)$ and let $\beta(i,j) = (i,j+1)$. Then $\alpha\beta = \beta\alpha$, α generates a cyclic subgroup π of order p, β generates a cyclic subgroup ρ of order p, and $\sigma = \pi \times \rho$ is a subgroup of $S(p^2)$ of order p^2.

Let W be a π-free acyclic complex and let ρ act on W through the isomorphism sending β into α. Then $W \otimes W$ is a $(\pi\times\rho)$-free acyclic complex by letting π act on the first factor and ρ on the second.

Let $Z_p^{(q)}$ denote the $S(p^2)$-module which is Z_p as an abelian group, and where a permutation acts by its sign if q is odd and trivially otherwise. Let R be any subgroup of $S(p^2)$. Let V be an R-free cyclic complex. By VII §2 we have a map

$$P': \ H^q(K;Z_p) \longrightarrow H_R^{p^2q}(V \times K^{p^2};Z_p^{(q)}) .$$

If R is a subgroup of σ, then $Z_p^{(q)}$ is a trivial R-module, since either $p = 2$ or R contains only even permutations.

Let $W_1 = W$ with π acting and let $W_2 = W$ with ρ acting. Then an action of $\pi \times \rho$ on $W_1 \times (W_2 \times K^p)^p$ can be defined by

$$(\alpha,\beta)(x \times (y_1 \times z_1) \times \ldots \times (y_p \times z_p)) =$$
$$= \alpha x \times (\beta y_{\alpha(1)} \times \beta z_{\alpha(1)}) \times \ldots \times (\beta y_{\alpha(p)} \times \beta z_{\alpha(p)})$$

for all $\alpha \in \pi$, $\beta \in \rho$, $x \in W_1$, $y_i \in W_2$, $z_i \in K^p$ (we regard both π and ρ as groups of cyclic permutations of p elements). We define an action of $\pi \times \rho$ on $W_1 \times W_2{}^p \times (K^p)^p$ by

$$(\alpha,\beta)(x \times y_1 \times \ldots \times y_p \times z_1 \times \ldots \times z_p) =$$
$$= \alpha x \times \beta y_{\alpha(1)} \times \ldots \times \beta y_{\alpha(p)} \times \beta z_{\alpha(1)} \times \ldots \times \beta z_{\alpha(p)}$$

where the variables have the same meaning as in the previous equation.

Now $W_1 \times W_2{}^p$ is a $(\pi \times \rho)$-free acyclic complex. Therefore we have the isomorphisms

$$\begin{aligned} H^*_{\pi\times\rho}(W_1 \times W_2 \times K^{p^2};Z_p) &\approx H^*_{\pi\times\rho}(W_1 \times W_2{}^p \times (K^p)^p;Z_p) \\ &\approx H^*_{\pi\times\rho}(W_1 \times (W_2 \times K^p)^p;Z_p) \\ &\approx H^*(W_1 \times_\pi (W_2 \times {}_\rho K^p)^p;Z_p) , \end{aligned}$$

where $\pi \times \rho$ acts on $K^{p^2} = (K^p)^p$ by

$$(\alpha,\beta)(z_1 \times \ldots \times z_p) = \beta z_{\alpha(1)} \times \ldots \times \beta z_{\alpha(p)} .$$

We therefore have an isomorphism

$$d_2^*\colon \ H^*(W_1 \times {}_\pi(W_2 \times {}_\rho K^p)^p;Z_p) \longrightarrow H^*(W_1 \times {}_\pi W_2 \times {}_\rho K^{p^2};Z_p)$$

which is induced by the diagonal $d_2\colon\ W_2 \longrightarrow W_2{}^p$.

1.1. LEMMA. The following diagram is commutative

$$\begin{array}{ccccc}
H^{pq}(K;Z_p) & \xrightarrow{\quad P \quad} & H^{pq}(W_2\times_\rho K^p;Z_p) & \xrightarrow{\quad d^* \quad} & H^{pq}(W_2/\rho\times K;Z_p) \\
\downarrow P' & & \downarrow P & & \downarrow P \\
H^{p^2q}(W_1\times_\pi W_2\times_\rho K^{p^2};Z_p) & \xleftarrow{(d_2)^*} & H^{p^2q}(W_1\times_\pi(W_2\times_\rho K^p)^p;Z_p) & \xrightarrow{d^*} & H^{p^2q}(W_1\times_\pi(W_2/\rho\times K)^p;Z_p) \\
\downarrow (d')^* & & \downarrow d_3{}^* & & \downarrow (d_2\times d)^* \\
H^{p^2q}(W_1/\pi\times W_2/\rho\times K;Z_p) & \xleftarrow{d^*} & H^{p^2q}(W_1/\pi\times(W_2\times_\rho K^p) & \xrightarrow{d^*} & H^{p^2q}(W_1/\pi\times W_2/\rho\times K;Z_p)
\end{array}$$

where $d'\colon\ K \longrightarrow K^{p^2}$, $d\colon\ K \longrightarrow K^p$

and $d_3\colon\ W_2 \times K^p \longrightarrow (W_2 \times K^p)^p$ are diagonals.

PROOF. The commutativity of the lower two squares follows since the maps of cohomology groups are induced by continuous maps which commute. The upper right hand square commutes because of VII 2.3. The upper left hand square commutes on the cochain level.

REMARK. To be quite rigorous one should point out that P was only defined on finite regular cell complexes (Chapter VII §2), while $W_2 \times_\rho K^p$ is certainly not finite, and may not be regular. We can ensure that $W_2 \times_\rho K^p$ is regular by replacing W_2 by its first derived. To make $W_2 \times_\rho K^p$ finite, we insist that W_2 should have a finite n-skeleton for each n (for example the complex of V §5), and then replace W_2 by its n-skeleton for some n much larger than p^2q.

By the Künneth theorem we can write

$$(d')^* P'u = \Sigma_{j,k}\, w_j \times w_k \times D_{j,k}u .$$

1.2. COROLLARY. $\Sigma_{j,k}\, w_j \times w_k \times D_{j,k}u = \Sigma_j\, w_j \times D_j(\Sigma_\ell\, w_\ell \times D_\ell u)$.

PROOF. From 1.1 we see that $(d')^*P' = (d_2 \times d)^*\, P\, d^*\, P$.

1.3. LEMMA. If $u \in H^q(K;Z_p)$, then

$$D_{j,k}u = (-1)^{jk+p(p-1)q/2}\, D_{k,j}u .$$

PROOF. Let $\lambda \in S(p^2)$ be the element such that $\lambda(i,j) = (j,i)$. Let λ^* denote the automorphisms induced by λ on the cohomology level (see V 3.4). Let V be an $S(p^2)$-free acyclic complex. Let $\sigma = \pi \times \rho$. Then by VII 2.5, V 3.3 and V 3.4 we have the commutative diagram

$$\begin{array}{ccccc} & \nearrow^{(d')^*P} & H_\sigma^{p^2q}(W{\times}W{\times}K;Z_p^{(q)}) & \xrightarrow{\lambda^*} & H_\sigma^{p^2q}(W{\times}W{\times}K;Z_p^{(q)}) \\ H^q(K;Z_p) & & \uparrow & & \uparrow \\ & \searrow_{(d')^*P} & H_{S(p^2)}^{p^2q}(V \times K;Z_p^{(q)}) & \xrightarrow{\lambda^*=1} & H_{S(p^2)}^{p^2q}(V \times K;Z_p^{(q)}) \end{array}$$

In order to determine the upper map λ^*, we have by V 3.4 to construct a chain map

$$\lambda_\#: W \otimes W \longrightarrow W \otimes W$$

such that $\lambda_{\#}\alpha = \beta\lambda_{\#}$ and $\lambda_{\#}\beta = \alpha\lambda_{\#}$ where α generates π and acts on the first factor and β generates ρ and acts on the second factor. Such a map is given by

$$\lambda_{\#}(v_1 \otimes v_2) = (-1)^{jk}(v_2 \otimes v_1)$$

where $\dim v_1 = j$ and $\dim v_2 = k$. Now λ transposes a $p \times p$ matrix and therefore it is a permutation with sign $(-1)^{p(p-1)/2}$. By V 1.2, $\lambda^*(w_j \times w_k \times D_{j,k}u)$ is represented by the $(\pi\times\rho)$-equivariant cocycle

$$W \times W \times K \xrightarrow{\lambda_{\#} \otimes 1} W \otimes W \otimes K \xrightarrow{w_j \otimes w_k \otimes D_{j,k}u} Z_p \xrightarrow{(-1)^{p(p-1)q/2}} Z_p.$$

This cocycle is equal to

$$(-1)^{jk+p(p-1)q/2}\, w_k \otimes w_j \otimes D_{j,k}u .$$

By the commutative diagram the lemma follows.

The proof of the Adem relations will be slightly simplified by the following conventions.

1.4. CONVENTION. $\binom{r}{j} = 0$ if $r < 0$ or $j < 0$; $\binom{r}{0} = 1$ if $r \geq 0$; $w_r \in H^r(\pi;Z_p)$ is zero if $r < 0$; Sq^j and P^j are zero for $j < 0$. All summations run from $-\infty$ to $+\infty$ unless otherwise stated.

By V 5.2 and I 2.4 we have $Sq^j w_r = \binom{r}{j} w_{r+j}$ and this now holds for all integers r and j. By V 5.2 and VI 2.2 we have

$$P^j w_{2r} = \binom{r}{j} w_{2r+2j(p-1)}.$$

By V 5.2 $\beta P^j w_{2r} = 0$. By the Cartan formula, V 5.2 and VI 2.2,

$$P^j w_{2r-1} = \binom{r-1}{j} w_{2r+2j(p-1)-1}$$

and $\beta P^j w_{2r-1} = -\binom{r-1}{j} w_{2r+2j(p-1)}$.

1.5. THEOREM. The Sq^i defined in VII 6.1 satisfy the Adem relations.

PROOF. If $\dim u = q$, we have

$$d^*Pu = \Sigma_i\, w_{q-i} \times Sq^i u.$$

By 1.2, 1.4 and the Cartan formula we have

$$(d')^*P'u = \Sigma_{i,k}\, w_{2q-k} \times Sq^k(w_{q-i} \times Sq^i u)$$
$$= \Sigma_{i,k,j}\binom{q-i}{j}\, w_{2q-k} \times w_{q-i+j} \times Sq^{k-j}Sq^i u\,.$$

Therefore

$$D_{2q-k,2q-\ell}u = \Sigma_i \binom{q-i}{q-\ell+i} Sq^{k+\ell-i-q}\, Sq^i u\,.$$

By 1.3,

$$D_{2q-k,2q-\ell}u = D_{2q-\ell,2q-k}u.$$

Therefore

$$\Sigma_i \binom{q-i}{q-\ell+i} Sq^{k+\ell-i-q}\, Sq^i u = \Sigma_r \binom{q-r}{q-k+r} Sq^{k+\ell-r-q}\, Sq^r u\,.$$

Let $q = 2^s - 1 + c$ and let $\ell = q + c$. The non-negative integers s,k and c are now arbitrary. Then

$$\binom{q-i}{q-\ell+i} = \binom{2^s-1+(c-i)}{(i-c)} = \begin{cases} 0 & \text{if } i \neq c \\ 1 & \text{if } i = c. \end{cases} \quad \text{by 1.4 and I 2.6,}$$

$$\binom{q-r}{q+r-k} = \binom{q-r}{k-2r} = \binom{2^s-1+c-r}{k-2r} \quad \text{since } \binom{x}{y} = \binom{x}{x-y}\,.$$

Now suppose that $k < 2c$. The binomial coefficient just examined is zero unless $2r \leq k$. Therefore it is zero unless $c - r > 0$. By I 2.6 this binomial coefficient is equal to $\binom{c-r-1}{k-2r}$ for $2^s > k$ and $r \geq 0$. Substituting in (1) we have that if $k < 2^s, 2c$ and $\dim u = q = 2^s-1 + c$, then (2) $Sq^k\, Sq^c u = \Sigma_r \binom{c-r-1}{k-2r} Sq^{k+c-r}\, Sq^r u$. By VII 6.8 the theorem is proved.

1.6. THEOREM. The P^i defined in VII 6.1 satisfy the Adem relations.

PROOF. By VII 3.5, VII 4.4 and VII 6.1, we have, writing $2m = p-1$ and $\nu(q) = (m!)^{-q}(-1)^{m(q^2+q)/2}$,

$$\nu(q)\, d^*Pu = \Sigma_i(-1)^i\, w_{(q-2i)2m} \times P^i u + \Sigma_i(-1)^i\, w_{(q-2i)2m-1} \times \beta P^i u\,.$$

By 1.2 we have

$$\nu(pq)\nu(q)(d')^*P'u = \Sigma_{k,i}(-1)^{i+k}\, w_{(pq-2k)2m} \times P^k(w_{(q-2i)2m} \times P^i u)$$
$$+ \Sigma_{k,i}(-1)^{i+k}\, w_{(pq-2k)2m} \times P^k(w_{(q-2i)2m-1} \times \beta P^i u)$$
$$+ \Sigma_{k,i}(-1)^{i+k}\, w_{(pq-2k)2m-1} \times \beta P^k(w_{(q-2i)2m} \times P^i u)$$
$$+ \Sigma_{k,i}(-1)^{i+k}\, w_{(pq-2k)2m-1} \times \beta P^k(w_{(q-2i)2m-1} \times \beta P^i u).$$

By the Cartan formula and 1.4 we have

$$P^k(w_{(q-2i)2m} \times P^i u) = \Sigma_j \binom{(q-2i)m}{j} w_{(q-2i+2j)2m} \times P^{k-j} P^i u ;$$

$$P^k(w_{(q-2i)2m-1} \times \beta P^i u) = \Sigma_j \binom{(q-2i)m-1}{j} w_{(q-2i+2j)2m-1} \times P^{k-j} \beta P^i u ;$$

$$\beta P^k(w_{(q-2i)2m} \times P^i u) = \Sigma_j \binom{(q-2i)m}{j} w_{(q-2i+2j)2m} \times \beta P^{k-j} P^i u ;$$

$$\beta P^k(w_{(q-2i)2m-1} \times \beta P^i u) = = \Sigma_j \binom{(q-2i)m-1}{j} w_{(q-2i+2j)2m} \times P^{k-j} \beta P^i u$$

$$- \Sigma_j \binom{(q-2i)m-1}{j} w_{(q-2i+2j)2m-1} \times \beta P^{k-j} \beta P^i u .$$

Therefore if $a = pq-2k$ and $b = q-2i+2j$, and the summations range over i,j,k, we have

(1) $\nu(pq)\nu(q)\, D_{2am,2bm}u = \Sigma_{i,j,k}(-1)^{i+k}\binom{(q-2i)m}{j} P^{k-j} P^i u ;$

(2) $\nu(pq)\nu(q)\, D_{2am,2bm-1}u = \Sigma_{i,j,k}(-1)^{i+k} \binom{(q-2i)m-1}{j} P^{k-j} \beta P^i u ;$

(3) $\nu(pq)\nu(q)\, D_{2am-1,2bm}u = \Sigma_{i,j,k}(-1)^{i+k} \binom{(q-2i)m}{j} \beta P^{k-j} P^i u$

$$- - \Sigma_{k,j,k}(-1)^{i+k}\binom{(q-2i)m-1}{j} P^{k-j} \beta P^i u .$$

Now $\nu(q)^4 \equiv 1 \pmod p$ by VII 6.3 and therefore $\nu(pq)\, \nu(q)$ has an inverse.

1.8. LEMMA. The first Adem relation is satisfied.

PROOF. Let $a = pq - 2k$ and $b = pq - 2\ell$. By 1.2 and (1) we have

(4) $\Sigma_i(-1)^{i+k}\binom{(q-2i)m}{mq-\ell+i} P^{k-mq+\ell-i} P^i u =$

$$= \Sigma_r(-1)^{r+\ell+mq}\binom{(q-2r)m}{mq-k+r} P^{\ell-mq+k-r} P^r u .$$

Let $q = 2(1 + \ldots + p^{s-1}) + 2c$ and let $\ell = c + mq$. The integers s,c and k are now arbitrary. Then

$$\binom{(q-2i)m}{mq-\ell+i} = \binom{(p-1)(1+ \ldots +p^{s-1}) + (p-1)(c-i)}{(i-c)}$$

$$= \begin{cases} 0 & \text{if } i \neq c \text{ by I 2.6 and 1.4} \\ 1 & \text{if } i = c. \end{cases}$$

Also

$$\binom{(q-2r)m}{mq-k+r} = \binom{(q-2r)m}{k-pr} \quad \text{since} \quad \binom{x}{y} = \binom{x}{x-y}$$

$$= \binom{p^s-1+(p-1)(c-r)}{k-pr} .$$

Now suppose that $k < pc$. The binomial coefficient just examined is zero unless $pr \leq k$. Therefore it is zero unless $r < c$. By I 2.6 this binomial coefficient is equal to $\binom{(p-1)(c-r)-1}{k-pr}$ for $p^s > k$ and $r \geq 0$.

Substituting in (4) we have that if $p^s > k < pc$ and $\dim u = q = 2(1+\ldots+p^{s-1}) + 2c$ then

$$P^k P^c u = \Sigma_r (-1)^{r+k} \binom{(p-1)(c-r)-1}{k-pr} P^{c+k-r} P^r u .$$

By VII 6.8 the lemma is proved.

1.9. LEMMA. The second Adem relation is satisfied.

PROOF. Let $a = (pq-2k)$ and $b = (pq-2\ell)$. By 1.3, (2) and (3) we have

$$(5) \qquad \Sigma_i (-1)^{i+k+mq+1} \binom{(q-2i)m-1}{mq+i-\ell} P^{k-mq-i+\ell} \beta P^i u$$

$$= \Sigma_r (-1)^{r+\ell+1} \binom{(q-2r)m}{mq+i-\ell} \beta P^{\ell-mq-r+k} P^r u$$

$$+ \Sigma_r (-1)^{r+\ell} \binom{(q-2r)m-1}{mq+r-k} P^{\ell-mq-r+k} \beta P^r u .$$

Let $q = 2p^s + 2c$ and let $\ell = c + mq$. The integers s, c and k are now arbitrary. Then

$$\binom{(q-2i)m-1}{mq-\ell+i} = \binom{(p-1)(1+\ldots+p^{s-1}) + (p-1)(c-i)}{i-c}$$

$$= \begin{cases} 0 & \text{if } i \neq c \\ 1 & \text{if } i = c . \end{cases} \quad \text{by I 2.6 and 1.4}$$

$$\binom{(q-2r)m}{mq-k+r} = \binom{(q-2r)m}{k-pr} = \binom{(p-1)(p^s+c-r)}{k-pr}$$

Now suppose that $k \leq pc$. The binomial coefficient just examined is zero unless $pr \leq k$. Therefore it is zero unless $r \leq c$. By I 2.6 this binomial coefficient is equal to $\binom{(p-1)(c-r)}{k-pr}$ for $p^s > k$ and $r \geq 0$. We also have

$$\binom{(q-2r)m-1}{mq-k+r} = \binom{(q-2r)m-1}{k-pr-1} = \binom{(p-1)(p^s+c-r)-1}{k-pr-1} .$$

This binomial coefficient is zero unless $pr < k$. Therefore it is zero unless $c < r$. By I 2.6 this binomial coefficient is equal to

$$\binom{(p-1)(c-r)-1}{k-pr-1} \quad \text{for} \quad p^s > k \quad \text{and} \quad r \geq 0.$$

Substituting in (5) we have that if $p^s > k \leq pc$ and $\dim u = q = 2p^s + 2c$ then

$$P^k \beta P^c u = \sum_r (-1)^{r+k} \binom{(p-1)(c-r)}{k-pr} \beta P^{c+k-r} P^r u$$

$$+ \sum_r (-1)^{r+k+1} \binom{(p-1)(c-r)-1}{k-pr-1} P^{c+k-r} \beta P^r u \quad .$$

By VII 6.8 the lemma is proved.

§2. Extensions to Other Cohomology Theories.

We now extend the definitions of P^i and Sq^i so that they operate on relative cohomology groups.

2.1. THEOREM. If F is a cohomology operation defined for absolute cohomology groups, then there is one and only one cohomology operation defined on both absolute and relative cohomology groups, which coincides with F on absolute cohomology. Furthermore these extensions to the relative groups of the reduced power operations Sq^i and P^i satisfy all the axioms (see I §1 and VI §1).

PROOF. If $a \in K$ we have a commutative diagram

$$\begin{array}{ccccccccc} 0 & \longrightarrow & H^q(K,a;G) & \longrightarrow & H^q(K;G) & \longrightarrow & H^q(a;G) & \longrightarrow & 0 \\ & & & & \downarrow F & & \downarrow F & & \\ 0 & \longrightarrow & H^r(K,a;G') & \longrightarrow & H^r(K;G') & \longrightarrow & H^r(a;G') & \longrightarrow & 0 \end{array}$$

By diagram chasing we obtain a unique definition for $F\colon H^q(K,a;G) \longrightarrow H^r(K,a;G')$. The definition is natural for maps of pairs where the second space is a point or is empty.

Let (K,A) be a pair of spaces. Let L be K with the cone on A attached. By excision we have an isomorphism

$$H^*(L,CA) \longrightarrow H^*(K,A).$$

Let c be the cone-point of CA. By the five lemma we have an isomorphism

$$H^*(L,CA) \longrightarrow H^*(L,c).$$

These constructions and isomorphisms are natural for mappings of pairs (K,A). Since we have defined F on $H^*(L,c)$, we obtain F on $H^*(K,A)$.

It is immediate to check that all the axioms listed in I §1 and VI §1 follows from the axioms for absolute cohomology. This proves the theorem.

We now have Sq^i and P^i defined on cohomology groups of pairs (K,L) where K is a finite regular cell complex and L is a subcomplex.

2.2. THEOREM. a) There is a unique definition of Sq^i and P^i on the singular cohomology groups of an arbitrary pair of spaces, which coincides with the definition on finite regular cell complexes.
b) There is a unique definition of Sq^i and P^i on the Cech cohomology groups of an arbitrary pair of spaces, which coincides with the definition already given on finite regular cell complexes.

The extensions in both a) and b) satisfy all the axioms in I §1 and VI §1.

PROOF. We shall leave the reader to check that the axioms are satisfied whenever we extend the definitions of Sq^i or P^i.

We first extend the definition to pairs (K,L) where L is an infinite regular cell complex and L a subcomplex. Now $H^q(K,L;Z_p)$ is naturally isomorphic to $\text{Hom}\,(H_q(K,L),Z_p)$. Therefore $H^*(K,L;Z_p)$ is the inverse limit of the groups $H^*(K_\alpha,L_\alpha;Z_p)$ where K_α and L_α vary over the finite subcomplexes of K and L. Since the reduced powers are natural this gives a unique definition on $H^*(K,L;Z_p)$. A continuous map from one pair of infinite complexes to another pair maps finite subcomplexes into subsets of finite subcomplexes. It follows that Sq^i and P^i are natural on the category of pairs (K,L) where K is a (finite or infinite) regular cell complex and L is a subcomplex.

Now we extend the definition to pairs (K,L) where K is a CW complex and L a subcomplex. According to J. H. C. Whitehead (see [2]), the pair (K,L) is homotopy equivalent to a pair of simplicial complexes. This obviously gives a unique and natural definition for P^i or Sq^i on $H^*(K,L)$.

We now give the definition on $H^*(X,Y)$, the singular cohomology of an arbitrary pair X,Y. Let SX and SY be the geometric realisations of the singular complexes of the spaces X and Y (see [2]). Then we have a singular homotopy equivalence $h: (SX,SY) \longrightarrow (X,Y)$. Moreover this map is natural for maps of pairs (X,Y). Since (SX,SY) is a pair of CW complexes, we have defined P^i and Sq^i in $H^*(SX,SY)$. Since

$$h^*: H^*(X,Y) \longrightarrow H^*(SX,SY)$$

is an isomorphism, this gives a unique and natural extension of P^i and Sq^i to singular cohomology groups. This proves the first part of the theorem.

We now extend Sq^i and P^i to Cech cohomology. The Cech cohomology groups of a pair (X,Y) are obtained by ordering the open coverings of (X,Y) according to whether one covering refines another, taking the nerves of the coverings, and then taking the direct limit of the cohomology groups of the nerves. Since we have introduced Sq^i and P^i into the cohomology structure of the nerve of each covering, and Sq^i and P^i are natural, this defines Sq^i and P^i uniquely on $H^*(X,Y)$. It is easy to see that Sq^i and P^i are natural with respect to continuous maps of pairs (X,Y). This completes the proof of the theorem.

§3. The Uniqueness Theorem.

In this section we shall prove that the Sq^i and the P^i are uniquely determined by the axioms 1)-5) in I §1 and 1)-5) in VI §1. We shall do this by investigating the cyclic product of spaces. We shall use Z_p as coefficients throughout this section.

3.1. LEMMA. Let K be a chain complex over Z_p. Then K is homotopically equivalent to the chain complex which is isomorphic to $H_*(K)$ as a graded module and has zero boundary.

PROOF. Let B_q be the boundaries in K and let D_q be a subspace of K_q which is complementary to the cycles. Then K is isomorphic to the complex which is $H_q(K) + B_q + D_q$ in dimension q, and whose boundary operator is zero on $H_q(K) + B_q$ and maps D_q isomorphically onto B_{q-1}. Therefore K is the direct sum of the chain complexes H and $(B + D)$. $B + D$ has the contracting homotopy s which is defined to be a map into D, which is zero on D_q and such that $s\colon B_q \longrightarrow D_{q+1}$ is the inverse of the boundary. We extend s to K by letting $s(H) = 0$. Let $\mu\colon K \longrightarrow H$ be the projection and let $\lambda\colon H \longrightarrow K$ be the injection. The $\mu\lambda = 1$ and $\lambda\mu \simeq 1$ by the homotopy s. This proves the lemma.

Let K and L be chain complexes. Let π be the cyclic group of order p acting by cyclic permutations on K^p and L^p. Let W be a π-free acyclic complex and let π act on $W \otimes K^p$ and $W \otimes L^p$ by the diagonal action.

3.2. LEMMA. If $f,g\colon K \longrightarrow L$ are chain homotopic, then

$$1 \otimes f^p, 1 \otimes g^p\colon \ W \otimes K^p \longrightarrow W \otimes L^p$$

are equivariantly homotopic.

PROOF. By VII 2.1 there is an equivariant map $h\colon I \otimes W \longrightarrow I^p \otimes W$ such that $h(\overline{0} \otimes w) = \overline{0}^p \otimes w$ and $h(\overline{1} \otimes w) = \overline{1}^p \otimes w$. Let $D\colon I \otimes K \longrightarrow L$ be the chain homotopy between f and g. Then we have the equivariant chain maps

$$I \otimes W \otimes K^p \xrightarrow{h\otimes 1} I^p \otimes W \otimes K^p \xrightarrow{\approx} W \otimes (I \otimes K)^p \xrightarrow{1\otimes D^p} W \otimes L^p .$$

The composition is the required equivariant chain homotopy.

3.3. COROLLARY. If $f\colon K \longrightarrow L$ is a homotopy equivalence, then $1 \otimes f^p$ is an equivariant homotopy equivalence.

From 3.1 and 3.3, we see that $W \otimes K^p$ and $W \otimes H^*(K)^p$ are equivariantly homotopy equivalent. Therefore $\mathrm{Hom}_\pi(W \otimes K^p, Z_p)$ is homotopy equivalent to $\mathrm{Hom}_\pi(W \otimes H_*(K)^p, Z_p)$.

We choose a direct sum splitting of $H_*(K)$ into components A_i, each isomorphic to Z_p. Then $H_*(K) = \Sigma_{i=1}^{\infty} A_i$. So

$$H_*(K)^p = \Sigma_{i=1}^{\infty} A_i{}^p + Z_p(\pi) \otimes B$$

where $B = \Sigma_{i_1 \leq i_2 \leq \cdots \leq i_p; i_1 < i_p} A_{i_1} \otimes \cdots \otimes A_{i_p}$.

The action of π on $A_i{}^p$ is by cyclic permutation and on $Z_p(\pi) \otimes B$ by the usual action on $Z_p(\pi)$ and the identity on B. So, if $H_*(K)$ is of finite type,

$$\mathrm{Hom}_\pi(W \otimes H_*(K)^p, Z_p) \approx \Sigma_i \mathrm{Hom}_\pi(W \otimes A_i{}^p, Z_p) + \mathrm{Hom}\,(W \otimes Z_p(\pi) \otimes B, Z_p) .$$

We then obtain immediately

3.4. LEMMA. Let K be a finite regular cell complex. Then, writing $(W \times K^p)/\pi = W \times_\pi K^p$,

$$H^*(W \otimes_\pi K^p) \approx \Sigma_i H^*(W/\pi \otimes A_i{}^p) + H^*_\pi(W \otimes Z_p(\pi) \otimes B) .$$

Let $W/\pi \times K$ be embedded in $W \times_\pi K^p$ by the diagonal map $d\colon K \longrightarrow K^p$.

3.5. REMARK. For any pair of spaces (X,A), $H^*(X)$, $H^*(A)$ and $H^*(X,A)$ are modules over $H^*(X)$ in an obvious way. Moreover it is easy to see that the maps in the cohomology sequence are consistent with the module structure. If we have a map $X \longrightarrow Y$, then the cohomology sequence of (X,A) gets an $H^*(Y)$ structure via the induced map $H^*(Y) \longrightarrow H^*(X)$. The cohomology sequence of $(W \times_\pi K^p, W/\pi \times K)$ is a module over $H^*(\pi)$ via the projection $W \times_\pi K^p \longrightarrow W/\pi$. The action of a class

$$u \in H^*(\pi) = H^*(W/\pi)$$

is multiplication by $u \times 1^p$, where 1 is the unit class (or augmentation) on K.

We have the maps

$$H^q(K) \xrightarrow{P} H^{pq}(W \times_\pi K^p) \xrightarrow{d^*} H^{pq}(W/\pi \times K) .$$

3.6. PROPOSITION. The image of d^* is the $H^*(\pi)$-module generated by the image of d^*P.

PROOF. By VII 4.1 it will be sufficient to show that $H^*_\pi(W \times K^p; Z_p)$ is the sum of $\mathrm{Im}\,\tau$, where τ is the transfer, and the $H^*(\pi)$-module generated by Im P. We see from 3.4 that we need only show

1) $H^*_\pi(W \otimes Z_p(\pi) \otimes B) \subset \mathrm{Im}\,\tau$ and
2) $H^*(W/\pi \otimes A_i{}^p)$ is generated as an $H^*(\pi)$-module by the element Pu_i, where u_i is dual to a generator of A_i.

PROOF of 1). Let $\lambda: Z_p \longrightarrow Z_p(\pi)$ be the map which sends $1 \in Z_p$ to $1 \in \pi$. Let $\nu: Z_p(\pi) \longrightarrow Z_p$ send $1 \in \pi$ to $1 \in Z_p$ and all other elements of π to zero. λ induces a map

$$\lambda^*: C^*(W \otimes Z_p(\pi) \otimes B; Z_p) \longrightarrow C^*(W \otimes Z_p \otimes B; Z_p) .$$

An equivariant cochain of $W \otimes Z_p(\pi) \otimes B$ is determined by its image under λ^*. We also have a map

$$\nu^*: C^*(W \otimes Z_p \otimes B; Z_p) \longrightarrow C^*(W \otimes Z_p(\pi) \otimes B; Z_p)$$

induced by ν. Since $\nu\lambda = 1$, it follows that $\lambda^*\nu^* = 1$.

We must show that any equivariant cocycle u in $W \otimes Z_p(\pi) \otimes B$ is the transfer of a cocycle in $W \otimes Z_p(\pi) \otimes B$. Now $\nu^*\lambda^*u$ is a cocycle on $W \otimes Z_p(\pi) \otimes B$. In order to prove that $\tau\nu^*\lambda^*u = u$, we need only show that $\lambda^*(\tau\nu^*\lambda^*u) = \lambda^*u$ since an equivariant cochain is determined by its image under λ^*. From the definition of τ, it follows that $\lambda^*\tau\nu^* = 1$. This proves 1).

PROOF of 2). Let $C_i = \mathrm{Hom}(A_i, Z_p)$ and let u_i generate C_i. The $H^*(\pi)$-structure on $H^*_\pi(W \times K^p)$ is given by cup-products with elements $v \times 1^p$ where $v \in H^*(\pi)$ (see 3.5). Therefore $H^*_\pi(W \otimes A_i{}^p) \approx H^*(\pi) \otimes C_i{}^p$ is a module over $H^*(\pi)$ generated by $1 \times u_i{}^p$. Now as in 3.1 we can consider A_i as a subspace of K_q for some q. Let L_q be a complementary space to A_i in K_q. We can represent u_i as a cochain by insisting that $u_i(L_q) = 0$. Now Pu_i is defined by the composition

$$W \otimes K^p \xrightarrow{\varepsilon \otimes 1} K^p \xrightarrow{(u_i)^p} Z_p$$

which is equal to $1 \otimes (u_i)^p$. Therefore $1 \otimes u_i{}^p = Pu_i$. This proves 2) and the proposition follows.

We now define a graded module $S = \{S^r\}$ where $S^r \subset H^r(W/\pi \times K)$ defined by the formula

$$S^r = \Sigma_{0 \leq j < (p-1)r/p} H^j(W/\pi) \otimes H^{r-j}(K).$$

Note that $j = pj - (p-1)j < (p-1)(r-j)$.

3.7. LEMMA. $\delta: H^r(W/\pi \times K) \longrightarrow H^{r+1}(W \times_\pi K^p, W/\pi \dot{\times} K)$ maps S^r monomorphically.

PROOF. By the cohomology exact seuqence, we need only show that $S^r \cap \operatorname{Im} i^* = 0$. By 3.6 we need only show that $S^r \cap \{H^*(\pi)\text{-module generated by } \operatorname{Im} d^*P\} = 0$. Now $d^*Pu = \Sigma_{j=0}^{q(p-1)} w_j \times D_j u$ by VII 3.2 and VII 4.4. By V 5.2 and VII 5.4

$$w_k\, d^*Pu = w_{k+q(p-1)} \times a_q u + \text{other terms.}$$

But $k + q(p-1) \geq (p-1)q$ which proves the lemma.

We now define a modified transfer τ' such that the following diagram is commutative.

$$\begin{array}{ccc} H^*(W \times K^p) & \xrightarrow{\ \tau'\ } & H^*(W \times_\pi K^p,\ W \times_\pi K_d) \\ & \searrow^{\tau} \qquad \swarrow_{j^*} & \\ & H^*(W \times_\pi K^p) & \end{array}$$

Let K^p be subdivided so that the triangulation is invariant under π and has the diagonal as a subcomplex (subdivide K to get a simplicial complex and then take the Cartesian product of the triangulation as defined in [3] p. 67). Since $H^*(W \times K^p) = H^*(K^p)$, we can represent any cohomology class of $W \times K^p$ by $\varepsilon \otimes u$ where u is a cocycle on K^p. If $w \in W$ and $\sigma \in K^p$, we define

$$\begin{aligned} \tau'(\varepsilon \otimes u)(w \otimes \sigma) &= \Sigma_{\alpha\in\pi}\ \varepsilon(\alpha w)\, u(\alpha\sigma) \\ &= \Sigma_{\alpha\in\pi}\ \varepsilon(w)\, u(\alpha\sigma) \end{aligned}$$

If $\sigma \in K_d$ then $\alpha\sigma = \sigma$ for all $\alpha \in \pi$ and so

$$\tau'(\varepsilon \otimes u)\,.\,(w \otimes \sigma) = p\,\varepsilon(w)\,u(\sigma) = 0\ .$$

Therefore $\tau'(\varepsilon \otimes u) \in C^*_\pi(W \times K^p, W \times K_d)$. This defines the modified transfer.

Let $h\colon K^p \longrightarrow K$ be the projection onto the first factor. Let

$$\delta\colon H^*(W/\pi \times K) \longrightarrow H^*(W \times_\pi K^p, W/\pi \times K).$$

Recall from 3.5 that δ is a homomorphism of $H^*(\pi)$-modules. Let $u \in H^q(K)$.

3.8. LEMMA. $-\delta(w_{2i-1} \times u) = w_{2i}\,.\,\tau'(1 \times h^*u)\ .$

If $p = 2$, $\qquad \delta(w_i \times u) = w_{i+1}\,.\,\tau'(1 \times h^*u)\ .$

PROOF. $w_i \cdot \tau'(1 \times h^* u)$ is given by (see p. 67 for definition of N)

$$W \otimes K^p \longrightarrow W \otimes K^p \otimes W \otimes K^p \xrightarrow{1 \otimes \varepsilon^p \otimes \varepsilon \otimes 1} W \otimes K^p \xrightarrow{1 \otimes N} W \otimes K^p \xrightarrow{w_i \otimes h} K \xrightarrow{u} Z_p .$$

The composition of the first two maps is equivariantly homotopic to the identity by V 2.2. Therefore $w_i \cdot \tau'(1 \times h^* u)$ is represented by

$$W \otimes K^p \xrightarrow{(-1)^{iq} w_i \otimes uhN} Z_p .$$

Let $h^{\#}u\colon K^p \longrightarrow Z_p$ be written as $u' + u''$ where $u' = 0$ on K_d and $u'' = 0$ on $K^p - K_d$. Then u'' is invariant since K_d is fixed under π. Therefore $u''N = 0$. Therefore $w_i \cdot \tau'(1 \times h^* u)$ is represented by

$$W \otimes K^p \xrightarrow{(-1)^{iq} w_i \otimes u'N} Z_p .$$

Now $u'' | K_d = u | K$. Therefore $\delta(w_j \times u)$ is given by

$$W \otimes_\pi K^p \xrightarrow{(-1)^{q+j} \partial} W \otimes_\pi K^p \xrightarrow{w_j \otimes u''} Z_p .$$

Now $\partial = \partial \otimes 1 + 1 \otimes \partial$ and we can leave out $\partial \otimes 1$ since w_j is a cocycle. Therefore $\delta(w_j \times u)$ is represented by

$$(-1)^{q+j}(w_j \otimes u'')(1 \otimes \partial) = (-1)^j (w_j \otimes \delta u'') .$$

Now $\delta u'' = \delta(h^{\#}u - u') = -\delta u'$ since u is a cocycle. Therefore

$$\delta(w_j \times u) = (w_j \times \delta u')(-1)^{j+1} .$$

We must show that for i even or $p = 2$, $-w_i \otimes u'N$ and $w_{i-1} \otimes \delta u'$ have the same class in $H^*(W \times_\pi K^p, W/\pi \times K)$. It is sufficient to show that these two cocycles have the same value on every relative cycle in $(W \times_\pi K^p, W/\pi \times K)$. Such a relative cycle has the form

$$\Sigma_{j=0}^{q+i} e_j \otimes_\pi c_{q-j+i}$$

where

$$\partial(\Sigma_{j=0}^{q+i} e_j \otimes_\pi c_{q-j+i}$$

Therefore for j even or $p = 2$

$$-e_{j-1} \otimes_\pi \partial c_{q-j+i+1} + N e_{j-1} \otimes_\pi c_{q-j+i} \in W \otimes_\pi K_d .$$

Therefore

$$N\, c_{q-j+i} - \partial\, c_{q-j+i+1} \in K_d .$$

Now since $u' = 0$ on K_d we have for i even or $p = 2$

$$(w_i \otimes u\, N)(\Sigma_{j=0}^{q+i} e_j \otimes c_{q-j+i}) = u\, Nc_q = u\, \partial c_{q+1}$$

$$= (-1)^q(\delta u)c_{q+1}$$

$$= -(w_{i-1} \otimes \delta u)(\Sigma_{j=0}^{q+i} e_j \otimes c_{q-j+i}) .$$

3.9. THEOREM. For a fixed odd prime p, the axioms 1 through 5 of VI §1 characterize the operations P^i $(i = 0,1,2\ldots)$. Precisely, if B^i $(i = 0,1,2\ldots)$ is any sequence of cohomology operations satisfying these axioms then, for each i, $B^i = P^i$.

PROOF. From the axioms we deduce that

(1) $\delta P^i = P^i\delta$ as in I 1.2.

(2) $P^i w_1 = 0$ from VI 2.2 and so $P^i(w_1 u) = w_1 P^i u$ by the Cartan formula.

(3) $$\Sigma_{i=0}^{k}(-1)^i w_{2i(p-1)}\, P^{k-i}(w_2 u) = \Sigma_{i=0}^{k}(-1)^i w_{2i(p-1)+2}\, P^{k-i}u + \Sigma_{i=0}^{k-1}(-1)^i w_{2(i+1)(p-1)+2}\, P^{k-i-1}u = w_2 \cdot P^k u .$$

(4) By 3.7, $\delta\colon H^r(W/\pi \times K) \longrightarrow H^{r+1}(W \times_\pi K^p, W/\pi \times K)$ maps S^r monomorphically.

(5) By 3.8, $-\delta(w_{2i-1} \times u) = w_{2i} \cdot \tau'(1 \times h^* u)$.

Let $\gamma = \Sigma_{i=0}^{k}(-1)^i w_{2i(p-1)+1} \times P^{k-i}u \in H^*(W/\pi \times K)$. We recall that δ is an $H^*(\pi)$-homomorphism by 3.5. We see that

$$\delta\gamma = \Sigma_{i=0}^{k}(-1)^i\, w_{2i(p-1)}\, \delta(w_1 P^{k-i}u).$$

$$= \Sigma_{i=0}^{k}(-1)^i\, w_{2i(p-1)}\, P^{k-i}(\delta(w_1 u)) \quad \text{by (1) and (2)}$$

$$= \Sigma_{i=0}^{k}(-1)^{i+1} w_{2i(p-1)}\, P^{k-i}(w_2 \cdot \tau'(1 \times h^* u)) \quad \text{by (5)}$$

$$= -w_2 P^k\, \tau'(1 \times h^* u) \quad \text{by (3)}$$

If $q = \dim u = 2s$ or $2s+1$, we put $k = s+1$. Then $2k > q$ and $\dim \tau'(1 \times h^* u) = q$, so $P^k \tau'(1 \times h^* u) = 0$ and $\delta\gamma = 0$.

Suppose $\{B^i\}$ satisfy the same axioms as $\{P^i\}$. Then we can define γ' by replacing P^i with B^i. As above $\delta\gamma' = 0$. Therefore

$$\delta(\Sigma_{i=0}^{s}(-1)^i w_{2i(p-1)+1} \times (P^{s-i+1} - B^{s-i+1})u) = \delta(\gamma - \gamma') = 0.$$

The term $i = s + 1$ is omitted since $P^0 - B^0 = 1 - 1 = 0$. Now

$$\begin{aligned} \dim(\gamma - \gamma') &= 2i(p - 1) + 1 + q + 2(s - i + 1)(p - 1) \\ &= 2(s + 1)(p - 1) + q + 1 \\ &= \begin{cases} 2(s + 1)p & \text{if } q = 2s + 1 \\ 2(s + 1)p - 1 & \text{if } q = 2s. \end{cases} \end{aligned}$$

Therefore $(p - 1)\dim(\gamma - \gamma')/p$

$$= \begin{cases} 2(s + 1)(p - 1) = 2s(p - 1) + 2(p - 1) & \text{if } q = 2s+1 \\ 2(s + 1)(p - 1) - (p - 1)/p & \\ \quad = 2s(p - 1) + 2(p - 1) - (p - 1)/p & \text{if } q = 2s. \end{cases}$$

Therefore $(p - 1)\dim(\gamma - \gamma')/p > 2i(p - 1) + 1 = \dim w_{2i(p-1)+1}$.

Therefore $(\gamma - \gamma') \in S^r$ where $r = \dim(\gamma - \gamma')$. Since $\delta(\gamma - \gamma') = 0$, 3.7 shows that $\gamma - \gamma' = 0$. Therefore $P^i u = B^i u$ for $0 \leq i \leq k$. If $i > k$ then $2i > 2k > \dim u$ and $P^i u = B^i u = 0$. The theorem is proved.

3.10. THEOREM. The axioms 1 through 5 characterize the operations Sq^i $(i = 0,1,2\ldots)$. Precisely if R^i $(i = 0,1,2\ldots)$ is any sequence of cohomology operations satisfying these axioms then, for each i, $R^i = Sq^i$.

PROOF. From the axioms we deduce

(1) $\delta Sq^i = Sq^i\delta$ as in I 1.2.

(2) $\Sigma_{i=0}^{k} w_i\, Sq^{k-i}(w_1 u) = \Sigma_{i=0}^{k} w_{i+1}\, Sq^{k-i}u + \Sigma_{i=0}^{k-1} w_{i+2}\, Sq^{k-i-1}u$

$$= w_1\, Sq^k u.$$

(3) $\delta\colon H^r(W/\pi \times K) \longrightarrow H^{r+1}(W \times_\pi K^p, W/\pi \times K)$ maps S^r monomorphically by 3.7.

(4) $\delta(1 \times u) = w_1\, \tau'(1 \times h^* u)$ by 3.8.

Let $\gamma = \Sigma_{i=0}^{k} w_i \times Sq^{k-i} u \in H^*(W/\pi \times K)$.

We recall that δ is an $H^*(\pi)$-homomorphism by 3.5. Therefore

$$\begin{aligned} \delta\gamma &= \Sigma_{i=0}^{k} w_i\, \delta(1 \times Sq^{k-i}u) \\ &= \Sigma_{i=0}^{k} w_i\, \delta Sq^{k-i}(1 \times u) \end{aligned}$$

$$= \Sigma_{i=0}^{k} w_i \ Sq^{k-i} \ \delta(1 \times u)$$

$$= \Sigma_{i=0}^{k} w_i \ Sq^{k-i}(w_1 \tau'(1 \times h^* u))$$

$$= w_1 \ Sq^k \tau'(1 \times h^* u) \ .$$

If $q = \dim u$, we put $k = q+1$. Then $Sq^k \tau'(1 \times h^* u) = 0$ and so $\delta\gamma = 0$.

Suppose $\{R^i\}$ satisfy the same axioms as $\{Sq^i\}$. Then we can define γ' by replacing Sq^i with R^i. As above $\delta\gamma' = 0$. Therefore

$$\delta(\Sigma_{i=0}^{q} w_i \times (Sq^{q+1-i} - R^{q+1-i})u) = \delta(\gamma - \gamma') = 0 \ .$$

Now $\dim (\gamma - \gamma') = 2q + 1$ and $i \leq q < (2q+1)/2$. Hence $\gamma - \gamma' \in S^r$. Therefore $\gamma - \gamma' = 0$ by 3.7. Therefore $Sq^i u = R^i u$ for $0 \leq i \leq k$. If $i > k$ then $Sq^i u = R^i u = 0$. So the theorem is proved.

BIBLIOGRAPHY

[1] J. Adem, "The relations on Steenrod powers of cohomology classes," Algebraic Geometry and Topology, Princeton 1957, pp. 191-238.

[2] J. Milnor, "On spaces having the homotopy type of a CW complex," Trans. A. M. S., 90 (1959), pp. 272-280.

[3] S. Eilenberg and N. E. Steenrod, "Foundations of Algebraic Topology," Princeton 1952.

APPENDIX

Algebraic Derivations of Certain Properties of The Steenrod Algebra

$\mathfrak{a}$, the Steenrod algebra mod p, has been defined in VI §2 (in I §2 for p = 2) in a purely algebraic manner as the free associative algebra $\underline{\mathfrak{a}}$ over Z_p generated by the elements P^i of degree $2i(p-1)$ and β of degree 1 (for p = 2 by Sq^i of degree i) modulo the ideal generated by β^2, $P^0 - 1$ ($Sq^0 - 1$ if $p = 2$) and the Adem relations. The theorem proved in this appendix (Theorem 2) is purely algebraic both in hypothesis and conclusion. It was proved in Chapters I, II and VI by allowing $\mathfrak{a}$ to operate on the cohomology groups of certain spaces. The proof to be given here will be purely algebraic. The only new step is an identity between binomial coefficients mod p which was proved by D. E. Cohen [1] in a paper on the Adem relations.

Let $P(\xi_1, \xi_2, \ldots)$ be the polynomial algebra over Z_p on generators ξ_i of degree $2(p^i - 1)$ (of degree $2^i - 1$ if $p = 2$). Let $E(\tau_0, \tau_1, \ldots)$ be the exterior algebra over Z_p on generators τ_i of degree $2p^i - 1$. Let $H = P \otimes E$ ($H = P$ if $p = 2$). We shall define a diagonal $\psi_H\colon H \longrightarrow H \otimes H$ which will make H a Hopf algebra. In doing so we are free to choose ψ_H on the generators ξ_i and τ_i and then ψ_H will be uniquely determined. Let

$$\psi_H \xi_i = \Sigma_i\, \xi_{k-i}^{p^i} \otimes \xi_i \quad \text{and} \quad \psi_H \tau_k = \tau_k \otimes 1 + \Sigma_i\, \xi_{k-i}^{p^i} \otimes \tau_i .$$

The following lemma is easily verified.

LEMMA 1. ψ_H is associative. H is a commutative associative Hopf algebra with an associative diagonal. H is of finite type.

From now on we shall give no special discussion of the case $p = 2$, since this can be obtained by replacing P^i with Sq^i and suppressing all arguments involving β or τ_i.

We define a homomorphism of algebras

$$\eta: \underline{\mathfrak{A}} \longrightarrow H^*$$

by letting $\eta(P^i)$ be the dual of $\xi_1{}^i$ and $\eta(\beta)$ the dual of τ_0 in the basis of admissible monomials.

THEOREM 2. The map η induces an epimorphism $\mathfrak{A} \longrightarrow H^*$ which sends no non-zero sum of admissible monomials to zero.

Theorem 2 has the following corollary.

THEOREM 3. a) η induces an isomorphism $\mathfrak{A} \longrightarrow H^*$. $\mathfrak{A}$ has a basis consisting of the admissible monomials.

b) $\mathfrak{A}$ is a Hopf algebra with diagonal given by

$$\psi(P^i) = \Sigma_j P^j \otimes P^{i-j} \quad \text{and} \quad \psi(\beta) = \beta \otimes 1 + 1 \otimes \beta.$$

c) H is the Hopf algebra dual to $\mathfrak{A}$.

PROOF of THEOREM 3. As in VI 2.1 we see that the admissible monomials span $\mathfrak{A}$. They are linearly independent by Theorem 2. Part a) of the theorem follows. Part b) is proved by showing that

$$\varphi_H^* \, \eta(P^i) = \Sigma \, \eta(P^j) \otimes \eta(P^{i-j}) \quad \text{and}$$

$$\varphi_H^* \, \eta(\beta) = \eta(\beta) \otimes 1 + 1 \otimes \eta(\beta),$$

where φ_H is the multiplication in H. Part c) is trivial.

We shall now prove Theorem 2. The first step is to show that η is zero on β^2 and on the Adem relations. It is easy to see that $\eta(\beta^2) = 0$, since if x is a monomial in H, then

$$< \eta(\beta^2), x > = < \eta(\beta) \times \eta(\beta), \psi_H x > = 0$$

by inspection of the formula for $\psi_H x$. In order to see that η maps each Adem relation to zero we need a lemma.

LEMMA 4. (Cohen [1]) if $0 \leq c < pd$ then

$$\binom{c+d}{c} \equiv \Sigma_j \, (-1)^{c+j} \binom{c+d}{j} \binom{(d-j)(p-1)-1}{c-pj} \bmod p.$$

PROOF. The formal power series in a variable t with coefficients in Z_p form a commutative ring. A power series whose constant coefficient is non-zero has a unique inverse under multiplication. Let f be the element of the ring given by

$$f(t) = ((1+t)^{p-1} - t^p)^{c+d}/(1+t)^{c(p-1)+1} .$$

The lemma will be proved by expanding $f(t)$ in two different ways.

If we apply the binomial theorem to the numerator of $f(t)$ we obtain:

$$f(t) = \Sigma_j \binom{c+d}{j}(-1)^j t^{pj}(1+t)^{(d-j)(p-1)-1} .$$

Since $c - pj < p(d - j)$ the expansion of $(1+t)^{(d-j)(p-1)-1}$ will contain t^{c-pj} only if $j < d$, in which case the coefficient of t^{c-pj} is $\binom{(d-j)(p-1)-1}{c-pj}$. Therefore the coefficient of t^c in $f(t)$ is

$$\Sigma (-1)^j \binom{c+d}{j}\binom{(d-j)(p-1)-1}{c-pj} .$$

On the other hand $(1+t)^p = 1 + t^p$ and so

$$(1+t)^{p-1} - t^p = 1 - t(1+t)^{p-1} .$$

Therefore

$$f(t) = (1 - t(1+t)^{p-1})^{c+d}/(1+t)^{c(p-1)+1}$$

$$= \Sigma_j (-1)^j \binom{c+d}{j} t^j (1+t)^{(j-c)(p-1)-1} .$$

If we expand $(1+t)^{(j-c)(p-1)-1}$ we obtain a term of the form t^{c-j} only if $j \leq c$. Let λ_{c-j} be the coefficient of this term. Then the coefficient of t^c in $f(t)$ is

$$\Sigma_j (-1)^j \lambda_{c-j} \binom{c+d}{j} .$$

Now $\lambda_0 = 1$ and the lemma will follow if $\lambda_k = 0$ for $k > 0$.

$$\lambda_k = \frac{(-(p-1)k-1)(-(p-1)k-2)\ \ldots .(-(p-1)k-1-k+1)}{k\ !}$$

$$= \pm \binom{pk}{k}$$

$$= 0 \text{ if } k > 0 \text{ by I 2.6.}$$

PROPOSITION 5. The homomorphism $\eta: \mathfrak{A} \longrightarrow H^*$ sends each Adem relation to zero.

PROOF. Suppose $x \in H$ is a monomial then

$$< P^{\alpha}P^{\beta}, x > \; = \; < P^{\alpha} \otimes P^{\beta}, \psi_H x > .$$

This is zero unless $x = \xi_1^j \xi_2^k$ and (for dimensional reasons) $\alpha + \beta = j + k(p + 1)$. Then

$$\begin{aligned} < P^{\alpha}P^{\beta}, \xi_1^j \xi_2^k > &= < P^{\alpha} \otimes P^{\beta}, (\xi_1 \otimes \xi_0 + \xi_0 \otimes \xi_1)^j (\xi_1^p \otimes \xi_1)^k > \\ &= < P^{\alpha} \otimes P^{\beta}, \Sigma_m \binom{j}{m} \xi_1{}^{m+pk} \otimes \xi_1{}^{j+k-m} > \\ &= \binom{j}{\alpha - pk} . \end{aligned}$$

Let $R(a,b) = -P^a P^b + \Sigma_i (-1)^{a+i} \binom{(b-i)(p-1)-1}{a-pi} P^{a+b-i} P^i$.

Then $R(a,b) = 0$ in $\mathfrak{A}$ if $a < pb$. Now $\eta\, R(a,b)$ could only be non-zero on monomials of the form $\xi_1^j \xi_2^k$ where $a + b = j + k(p+1)$ and its value on such monomials is

$$-\binom{a+b-k(p+1)}{a-pk} + \Sigma_i (-1)^{a+i} \binom{(b-i)(p-1)-1}{a-pi} \binom{a+b-k(p+1)}{a+b-i-pk} .$$

By Lemma 4 with $d = b - k$, $c = a - pk$ and $j = i - k$, this expression is zero mod p.

We now have to show that if $a \leq pb$, then η sends the following expression to zero

$$\text{(1)} \qquad - P^a \beta P^b + \Sigma_i (-1)^{a+i} \binom{(b-i)(p-1)}{a-pi} \beta P^{a+b-i} P^i + \Sigma_k (-1)^{a+i} \binom{(b-i)(p-1)-1}{a-pi-1} P^{a+b-i} \beta P^i.$$

Let $Q \in H^*$ be dual to $\tau_1 \in H$. Then

$$\text{(2)} \qquad \eta(P^a \beta - \beta P^a) = Q\, \eta(P^{a-1})$$

To see this we note that $\eta(P^a\beta), \eta(\beta P^a)$ and $Q\, \eta(P^{a-1})$ are zero except on monomials of the form $\xi_1^a \tau_0$ and $\xi_1^{a-1} \tau_1$, and on these monomials the identity (2) is easy to check.

If $a < pb$ then the expression (1) is sent to zero by η, as we see on using (2) and $\eta(R(a,b)) = 0$ and $\eta(R(a-1,b)) = 0$. If $a = pb$ then (1) becomes

$$- P^a \beta P^b + \beta P^a P^b .$$

Under η this becomes $-Q\,\eta(P^{pb-1}P^b)$. Now

$$\eta(P^{pb-1}P^b) = \eta(R(pb-1,b)) = 0 .$$

This proves the proposition.

COROLLARY 6. η induces a homomorphism of algebras $\eta\colon \mathfrak{A} \longrightarrow H^*$.

As in VI 4.2 we can set up a one-to-one correspondence between sequences $I = (\varepsilon_0, i_1, \varepsilon_1, \ldots, i_k, \varepsilon_k, 0, \ldots)$ with $\varepsilon_r = 0$ or 1 and $i_r = 0,1,2,\ldots$ and admissible sequences $I' = (\varepsilon_0, i_1, \varepsilon_1, \ldots, i_k', \varepsilon_k, 0, \ldots)$ by the equations

$$i_r = i_r' - pi_{r+1}' - \varepsilon_r .$$

Let $P^{I'} = \beta^{\varepsilon_0} P^{i_1'} \beta^{\varepsilon_1} \ldots P^{i_k'} \beta^{\varepsilon_k}$ and let

$$\xi^I = \tau_0^{\varepsilon_0} \xi_1^{i_1} \tau_1^{\varepsilon_1} \cdots \xi_k^{i_k} \tau_k^{\varepsilon_k} .$$

Then ξ^I and $P^{I'}$ have the same degree. We order the set of sequences $\{I\}$ lexicographically from the right.

LEMMA 7. $\langle P^{I'}, \xi^J \rangle$ is zero for $I < J$ and ± 1 for $I = J$.

PROOF. We prove this by induction on the degree of ξ^J. It is true in degree 0.

Case 1). The last non-zero element of I' is i_k'. Let M' be the sequence I' with i_k' replaced by 0. We have

$$i_k' = i_k \text{ and } M = (\varepsilon_0, i_1, \varepsilon_1, \ldots, i_{k-1} + pi_k, \varepsilon_{k-1}, 0 \ldots) .$$

(If $k = 1$, $M = (\varepsilon_0, 0, \ldots)$.)

Now $\langle P^{I'}, \xi^J \rangle = \langle P^{M'} \otimes P^{i_k}, \psi_H \xi^J \rangle$. By our induction hypothesis we need only take into account terms of the form $\xi^L \otimes \xi_1^{i_k}$ where $L \le M$ in the expansion of $\psi_H \xi^J$. Inspecting the formula for ψ_H we see that $\langle P^{I'}, \xi^J \rangle = 0$ unless J and I have the same length and $j_k \le i_k$. Since $J \ge I$, we can assume that $j_k = i_k$ and that J and I have the same length. Therefore in the expansion of $\psi_H \xi^J$ we need only take into account the term $\xi^L \otimes \xi_1^{i_k}$ where

$$L = (\delta_0, j_1, \delta_1, \ldots, j_{k-1} + pj_k, \delta_{k-1}, 0, \ldots) .$$

(If $k = 1$, $L = (\delta_0, 0, \ldots)$.)

So $L \ge M$ and we have $L = M$ if and only if $J = I$. By our induction hypothesis the lemma follows in this case.

Case 2). The last non-zero term of I' is ε_k. Let M' be the sequence I' with ε_k replaced by zero. Then

$$M = (\varepsilon_0, i_1, \varepsilon_1, \ldots, i_{k-1}, \varepsilon_{k-1}, i_k + 1, 0, \ldots).$$

(If $k = 0$, $M = (0,0,\ldots)$.)

Now

$$< P^{I'}, \xi^J > = < P^{M'} \otimes \beta, \psi_H \xi^J > .$$

By our induction hypothesis we need only take into account terms of the form $\xi^L \otimes \tau_0$ where $L \leq M$ in the expansion of $\psi_H \xi^J$. Inspecting the formula for ψ_H we see that $< P^{I'}, \xi^J > = 0$ unless J and I have the same length (and so $\delta_k = \varepsilon_k = 1$). We assume that J and I have the same length. Then in the expansion of $\psi_H \xi^J$ we need only take into account the term $\xi^L \times \tau_0$ where

$$L = (\delta_0, j_1, \delta_1, \ldots, j_{k-1}, \varepsilon_{k-1}, j_k + 1, 0, \ldots).$$

(If $k = 0$, $L = (0,0,\ldots)$.)

So $L \geq M$ and we have $L = M$ if and only if $J = I$. The lemma follows.

We now show that η is an epimorphism. On each degree there are only a finite number of monomials ξ^J. By a decreasing induction on J, and using Lemma 7, the image of η is seen to contain the dual of ξ^J. Moreover η does not send the sum of admissible monomials $\Sigma \lambda_i P^{I_i'}$ to zero, as we see by applying Lemma 7 to the term for which I_i is greatest. This proves Theorem 2.

BIBLIOGRAPHY

[1] D. E. Cohen, "On the Adem relations," Proc. Camb. Phil. Soc., 57 (1961) pp. 265-266.

INDEX

Errata for the Annals Study No. 50

COHOMOLOGY OPERATIONS

by

D. B. A. Epstein and N. E. Steenrod

On page 60, line 14, it should be assumed explicitly in the definition of a group π of automorphisms of a complex K that, if $\alpha \in \pi$ and σ is a cell of K such that $\alpha\sigma = \sigma$, then $\alpha | \sigma$ is the identity. The reason for the requirement is that $H^*_\pi(K)$ is not otherwise a topological invariant of $(\pi, |K|)$. In subsequent sections the condition holds automatically because π acts freely.

On page 68, the Bockstein operator β is introduced with a reversed sign. If we adhere to the boundary-coboundary formula (as was intended)

$$u \cdot \partial e = (-1)^{q+1} \delta u \cdot e, \qquad q = \dim u = \dim e - 1$$

then we obtain $\beta w_1 = w_2$ on lines 14, 18, 25, and on line 17 the three minus signs should be deleted. In accord with this, on page 107 line 3, insert a minus sign before $D_{2k-1}u$; on line 6 delete the sign, and on line 7 change the second minus to a plus.

On page 106, we may add to the statement of Lemma 4.5: When $p = 2$ and q is odd, then $\beta' d^* Pu = 0$ where β' is the twisted Bockstein associated with $0 \to Z_2 \to Z_4 \to Z_2 \to 0$ in which the generator T of π operates by reversing sign. In the subsequent proof observe that v^2 would not be equivariant if T did not reverse sign. The statement of Corollary 4.6 should now include: If $p = 2$ and q is odd, then $\beta D_{k+1} u = D_{2k} u$. Note that $\beta = \beta'$ on K. After these changes the proof on page 114 of Theorem 6.7 may be completed without appealing to 6.8.

On page 112, in the definition 6.1, replace $(m!)^q$ by $(m!)^{-q}$ where of course the inverse is taken in the field Z_p. By 6.3, this is at most a change in sign. This does not require any changes on page 113 since the correct formula was used there. However, on page 119, line -7, in the definition of $\nu(q)$ we must replace $(m!)^{-q}$ by $(m!)^q$.

Page	Line	This	should be	that.
1	-8	q		n
9	12	Sq^i		Sq^I
19	17	$m = M_k$		$m \neq M_k$
44	1	$G(n)$		$G(m)$
53	11	u		U
55	10	vectors		vector fields
62	12	cyclic		acyclic
67	-6	$i \leq j$		$i < j$
70	-1	$i \mid p-1$		$p-1 \mid i$
71	12	$\alpha u \cdot \alpha^{-1} c$		$\alpha(u \cdot \alpha^{-1} c)$
74	1	let ρ be		let p be
77	3	$a \leq b$		$a \leq pb$
84	5	$\tau(I')$		$\tau(I'))$
108	-4, -3	$\varepsilon \otimes d_{\#}, d_{\#}$		$(1 \times d)_{\#}$
109	13	formula		formula for
110	17	α_i (on left)		α_1
113	9	n		r
119	8	Insert the number (1) of this formula		
119	-7	$(m!)^{-q}$		$(m!)^q$
119	-2, -3	$(-1)^{i+k}$		$(-1)^{i+k+1}$
120	5	$= = \sum_j$		$= \sum_j$
120	8-12	In the three formulas, replace each j by $i - (q-b)/2$ and each sum is taken over all i (for which the binomials are not zero); k is fixed.		
120	12	$- - \sum$		$+ \sum$

On page 116, lines 14 and 18, the asserted isomorphism is only a homomorphism because $W_1 \times (W_2 \times K^p)^p$, when collapsed under $\pi \times \rho$, must be collapsed still more to obtain $W_1 \times_\pi (W_2 \times_\rho K^p)^p$. As a consequence the diagram 1.1 is not adequate to prove Corollary 1.2. Replace the argument from line 10 of page 116 to line 4, page 117 by the following.

Let $\rho_i \subset S(p^2)$ be the cyclic group of order p generated by β_i where $\beta_i(k, j) = (k, j)$ if $k \neq i$ and $\beta_i(i, j) = (i, j+1)$. Since $\beta_0, \ldots, \beta_{p-1}$ commute, the direct product $\rho_0 \times \rho_1 \ldots \times \rho_{p-1}$ is a subgroup of $S(p^2)$. Clearly ρ is the diagonal of $\rho_0 \times \ldots \times \rho_{p-1}$. Let τ be the subgroup generated by π and $\rho_0 \times \ldots \times \rho_{p-1}$. Since $\alpha\beta_i \alpha^{-1} = \beta_{i+1}$, τ is a split extension of $\rho_0 \times \ldots \times \rho_{p-1}$ by π, and π acts as cyclic permutations of the factors. Since the order of τ is p^{p+1}, it follows that τ is a Sylow p-group for $S(p^2)$. (It is notationally suggestive to write τ in the form $\pi * \rho^p$.)

Define the action of τ in $W_1 \times (W_2 \times K^p)^p$ in the following natural way. ρ_i acts as the identity in all factors except for the ith factor $W_2 \times K^p$ where β_i acts as does β, i.e., it acts as usual in W_2 and permutes the factors of K^p cyclically. π acts as usual in W_1 and permutes the p factors of the form $W_2 \times K^p$ cyclically. It is readily checked that this gives an action of τ consistent with the inclusion $\pi \times \rho \subset \tau$.

Let $h\colon W_1 \times W_2^p \times (K^p)^p \to W_1 \times (W_2 \times K^p)^p$ be the isomorphism which shuffles the W_2^p with $(K^p)^p$. Define the action of τ in $W_1 \times W_2^p \times (K^p)^p$ so that h is τ-equivariant. Then h induces an isomorphism

$$(W_1 \times W_2^p) \times_\tau K^{p^2} \to W_1 \times_\tau (W_2 \times K^p)^p$$

The complex on the right is readily seen to be isomorphic to $W_1 \times_\pi (W_2 \times_\rho K^p)^p$. Thus h induces an isomorphism

$$h^*\colon H^*(W_1 \times_\pi (W_2 \times_\rho K^p)^p) \approx H^*(W_1 \times W_2^p \times_\tau K^{p^2})$$

We obtain thereby the accompanying diagram.

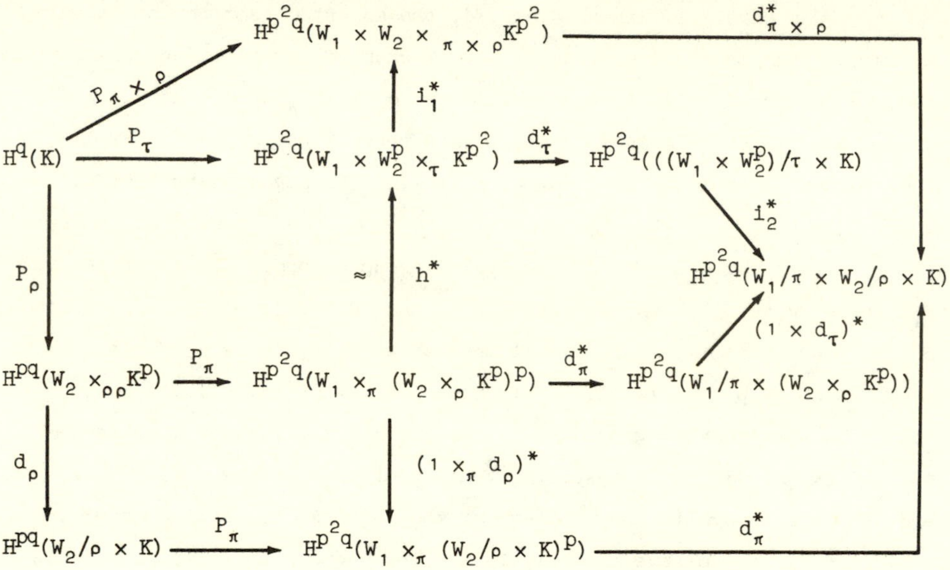

The mappings i_1^* and i_2^* are induced by the inclusion $\pi \times \rho \subset \tau$, the diagonal mapping $W_2 \to W_2^p$ and the identity mappings of W_1, K and K^{p^2}. On the right side of the diagram, commutativity holds because the corresponding diagram of mappings is commutative. By VII, 2.5, the triangle is commutative. By VII, 2.3, the lower square on the left is commutative. Finally the relation $h^* P_\pi P_\rho = P_\tau$ is proved by applying the definitions on the chain level. In both cases $W_1 \otimes W_2^p \otimes K^{p^2}$ is projected into K^{p^2}, and K^{p^2} is mapped into Z_p by u^{p^2}.

The commutativity of the diagram yields the formula

$$d^*_{\pi \times \rho}\ P_{\pi \times \rho} = (d_\pi^*\ P_\pi)(d_\rho^*\ P_\rho)$$

If we apply to this the Kunneth formula, we obtain Corollary 1.2.

ANNALS OF MATHEMATICS STUDIES

105. Multiple Integrals in the Calculus of Variations and Nonlinear Elliptical Systems, *by* MARIANO GIAQUINTA
104. Etale Homotopy of Simplicial Schemes, *by* ERIC M. FRIEDLANDER
103. Seminar on Minimal Submanifolds, edited *by* ENRICO BOMBIERI
102. Seminar on Differential Geometry, *edited by* SHING-TUNG YAU
101. Random Fourier Series With Applications to Harmonic Analysis, *by* MICHAEL B. MARCUS *and* GILLES PISIER
100. Recent Developments in Several Complex Variables, *edited by* JOHN E. FORNAESS
99. The Spectral Theory of Toeplitz Operators, *by* L. BOUTET DE MONVEL *and* V. GUILLEMIN
98. K-Theory of Forms, *by* ANTHONY BAK
97. Riemann Surfaces and Related Topics: Proceedings of the 1978 Stony Brook Conference, *edited by* IRWIN KRA *and* BERNARD MASKIT
96. Base Change for GL(2), *by* ROBERT P. LANGLANDS
95. C*-Algebra Extensions and K-Homology, *by* RONALD G. DOUGLAS
94. Continuous Cohomology, Discrete Subgroups, and Representations of Reductive Groups, *by* A. BOREL *and* N. WALLACH
93. Seminar on Micro-Local Analysis, *by* VICTOR W. GUILLEMIN, MASAKI KASHIWARA, *and* YAKAHIRO KAWAI
92. The Classifying Spaces for Surgery and Cobordism of Manifolds, *by* IB MADSEN *and* R. JAMES MILGRAM
91. Seminar on Singularities of Solutions of Linear Partial Differential Equations, *edited by* LARS HÖRMANDER
90. Infinite Loop Spaces, *by* J. F. ADAMS
89. Invariant Forms on Grassmann Manifolds, *by* WILHELM STOLL
88. Foundational Essays on Topological Manifolds, Smoothings, and Triangulations, *by* ROBION C. KIRBY *and* LAURENCE C. SIEBENMANN
87. Scattering Theory for Automorphic Functions, *by* PETER D. LAX *and* RALPH S. PHILLIPS
86. 3-Manifolds, *by* JOHN HEMPEL
85. Entire Holomorphic Mappings in One and Several Complex Variables, *by* PHILLIP A. GRIFFITHS
84. Knots, Groups, and 3-Manifolds, *edited by* L. P. NEUWIRTH
83. Automorphic Forms on Adele Groups, *by* STEPHEN S. GELBART
82. Braids, Links, and Mapping Class Groups, *by* JOAN S. BIRMAN
81. The Discrete Series of GL over a Finite Field, *by* GEORGE LUSZTIG
80. Smoothings of Piecewise Linear Manifolds, *by* M. W. HIRSCH *and* B. MAZUR
79. Discontinuous Groups and Riemann Surfaces, *edited by* LEON GREENBERG
78. Strong Rigidity of Locally Symmetric Spaces, *by* G. D. MOSTOW
77. Stable and Random Motions in Dynamical Systems: With Special Emphasis on Celestial Mechanics, *by* JÜRGEN MOSER
76. Characteristic Classes, *by* JOHN W. MILNOR *and* JAMES D. STASHEFF

A complete catalogue of Princeton mathematics and
science books, with prices, is available upon request.

PRINCETON UNIVERSITY PRESS
41 WILLIAM STREET, PRINCETON, NEW JERSEY 08540

A